D0312653

DATE DUE

Passionate Minds

Passionate Minds

The Inner World of Scientists

LEWIS WOLPERT

Professor of Biology as applied to Medicine,
University College London

and

ALISON RICHARDS

Producer, Science Programmes,
BBC Radio

A companion volume to
A Passion for Science

Oxford   New York   Tokyo

OXFORD UNIVERSITY PRESS

1997

Oxford University Press, Great Clarendon Street, Oxford OX2 6DP

Oxford   New York

Athens   Auckland   Bangkok   Bombay
Calcutta   Cape Town   Dar es Salaam   Delhi
Florence   Hong Kong   Istanbul   Karachi
Kuala Lumpur   Madras   Madrid   Melbourne
Mexico City   Nairobi   Paris   Singapore
Taipei   Tokyo   Toronto   Warsaw
and associated companies in
Berlin   Ibadan

By arrangement with BBC Books, a division of BBC Enterprises Ltd

Oxford is a trade mark of Oxford University Press

Published in the United States
by Oxford University Press Inc., New York

A catalogue record for this book is available from the British Library

Library of Congress Cataloging in Publication Data (Data available)

ISBN 0 19 854904 0

Typeset by Palimpsest Book Production Limited,
Polmont, Stirlingshire

Printed in Great Britain by
Bookcraft (Bath) Ltd
Midsomer Norton, Avon

Contents

EUREKA

REFLECTIONS

Acknowledgements

These interviews were originally conducted by Lewis Wolpert and produced by Alison Richards for BBC Radio 3. We wish to thank the scientists for permission to publish the interviews and for supplying photographs. We are also grateful to Maureen Maloney for all her help in the preparation of the manuscript.

CHAPTER 1

Passionate Minds

OUTSIDE their own habitat, scientists, as a species, are little understood. If they feature in popular awareness at all, it is through a limited set of media stereotypes. With a few exceptions, if scientists are not mad or bad, they are personality-free, their measured tones and formal reports implying ways of thinking and working far removed from the intellectual and emotional messiness of other human activities.

We embarked on these conversations, which were originally broadcast on BBC Radio 3, as an attempt to redress the balance. We wanted to give a rare glimpse of the intellectual, emotional and imaginative vigour that is the human reality of scientific life. Even a sample of this size demonstrates with extraordinary force the vitality and diversity of mind and temperament jostling within the retaining walls of 'science'. So vividly does the force of each personality spill out from the transcripts—and these can only hint at the voice and emotional colour of the original exchange—that they give the lie once and for all to the notion that science is in any way an 'inhuman' activity. Scientists think and feel about their work using the same psychological apparatus as the rest of us. It also becomes clear within a few pages that there is no one way of 'doing' science, even within recognized disciplines or groups. Among the experimentalists and the theorists, the biologists and the physicists, there are as many differences in style and motivation as there are in haircuts and accents.

These conversations show that it *is* possible for non-scientists to gain a meaningful sense of how scientists 'tick'; or at least as meaningful as the glimpse most of us get from listening to a poet or painter reflect upon their work. The dialogue is accessible, inspiring and full of surprises. We believe that these conversations make a case that society should—and can—be as interested in finding out about the life and times of Nobel scientists as in delving into the psyche of the latest literary prizewinners.

Such activities afford twin delights. Science—as we have just noted—seethes with diversity. Part of the fun is to explore and savour the uniqueness of each individual talent. On the other hand, it is also illuminating to seek patterns and underlying unities. We don't believe for a minute that there is a grand theory of scientific discovery any more than there is a grand theory of painting, but this collection of conversations also offers pleasurable scope for trend spotting amidst the infinite variety of genius.

For example, a casual glance through the following pages suggests that being an outsider is a definite help when it comes to making important scientific discoveries. The Spanish developmental biologist, Antonio Garcia Bellido, conducted much of his fundamental research in a country lacking an established scientific community or tradition. In the case of the evolutionary biologist, Richard Lewontin, it is his radical politics which set him apart: 'I feel alienated from [the scientific community]. I have little or nothing in common with my colleagues. We take the opposite point of view on almost everything . . . on every issue we struggle.'

In some cases the outsiders' cultural baggage plays a part in their success. The Polish-born poet and chemist Roald Hoffmann refers to being one of 'Hitler's last gifts to America'. The cell biologist Mike Berridge, who grew up in Rhodesia, cites the discipline of a colonial way of life. But being an outsider clearly confers a more fundamental advantage—notably the ability to view a subject or problem with an eye unclouded by history, habit or dogma. The Nobellist, Sir James Black invented beta-blockers and the ulcer drug, cimetidine, because he dared to enter a new field and didn't mind asking questions which were 'really quite preposterous'.

The immunologist and neurobiologist Gerald Edelman talks in terms of an 'innocence' which means you fail to realize how difficult a question it is that you are asking. Had he not been thus protected he would never have contemplated probing the structure of the antibody molecule. Smart people, as Richard Lewontin observes, 'are not crazy enough to spend their lives trying to solve really hard problems'.

Innocents and outsiders may also disregard—or fail to register—conventional barriers. The physicist and historian of science, Gerald Holton, makes the point that Freud, Einstein and Darwin didn't divide up the world in the conventional way: 'Think of Freud. The difference between a mature man or woman and a child . . . [is] obvious; [but] to Freud [it is] not at all obvious, he wants to find a continuity between what to us are completely separate things. The same with Darwin seeing the link between human species and other species.' Neither the particle physicist Carlo Rubbia, nor James Lovelock, originator of the Gaia theory, see science or nature as departmentalized. For Rubbia, 'There is only one type of science and the various fields are chapters of the same book'. In Lovelock's mind, 'The territories, the disciplines, are purely feudal, set up by professors to retain territories over which they have control'.

The ability to take advantage of this kind of intellectual free market is frequently cited as a key to scientific success. The physical anthropologist David Pilbeam sees it as his main strength, 'the kind of synthesis that occasionally opens up new questions and new ways of thinking about things. Putting material, diverse material, together in ways that haven't been done before.' Michael Berridge, who made fundamental discoveries about the way cells communicate with each other, is equally definite: 'If I was thinking of any single gift that one needs, it's this ability to make connections between a lot of disparate facts . . . The gifted scientists that I meet . . . have this facility. They have a very broad view of what's going on and they're able to make connections between different ideas, different disciplines.'

'Freelance personality' is the term that Rubbia uses to describe the kind of intellectual buccaneer that every successful scientist must be. But it is apparent that other qualities are essential, too. High on the list comes persistence, a concept that reverberates like an echo through the conversations. Scientists clearly find science hard. It's 'extremely difficult' says Mike Berridge, 'because you're up against nature . . . [it's] very much like a battle. You're like a general marshalling his forces to try and unlock some of the secrets.' Good ideas don't seem to come up very often: 'I think one's doing extremely well if one has a good idea once every, let's say, six months' observes the immunologist Avrion Mitchison, drily. Even then, of course, there's no guarantee that the idea will work. There may be insurmountable practical problems. Anne McLaren is a mammalian developmental biologist, whose doctoral project required pregnant rabbits, but 'none of the rabbits in the Oxford animal house got pregnant in the whole of the first year that I was supposed to be doing this Ph.D.' In the event she had to find a completely different problem. Even with the practical difficulties overcome, it may turn out that the brilliant idea you had was way off course. The late Nobel chemist, Peter Mitchell, recalled 'I had a particular view . . . and it turned out that at the end of nearly eight years I was wrong.' Moreover, this experience seems to be the rule rather than the exception. According to the inventor of the contraceptive pill, Carl Djerassi, '. . . scientific research most of the time is just a series of failures'. Another Nobellist, the physicist Murray Gell-Mann describes the constant uncertainty and worry that for him is an inescapable part of research: 'No matter what I've come up with, I've doubted it and worried about it and been terrified that it was wrong or trivial.' Science, asserts Mike Berridge, always involves 'long periods of gloom'.

Clearly, however, good scientists don't give up. One aspect of this seems to be their own certainty, whether in the short or long term, that they are on the right track. They have to have the talent, as Gerald Holton describes it, 'to smell out where it will all lead, before it becomes simple to show it'. It is having the ability to discern when it is the meter readings that are at fault rather than the idea that is wrong. Such people also have to have a hefty helping of courage and stubbornness to sustain their assault on the conventional wisdom. When Edelmen was working on the structure of antibodies 'People thought I was crazy—and although they didn't shy away from me at the lunch table, they thought that my work was pretty preposterous'. Sheldon Glashow, who shared the 1979 Nobel prize for physics for his work towards the unification of forces, describes how it was necessary to believe in quarks for years before 'these crazy ideas . . . these particles that no one had ever seen, and according to the theory today, no one ever will' were validated experimentally. 'If you would simply take all the kookiest ideas of the early 1970s and put them together, you would have made for yourself the theory which is, in fact, the correct theory of nature. So it was like madness, it was everybody's weirdest fancy was right.' For some it took even longer. It was the best part of twenty years before either Peter Mitchell's or James Lovelock's ideas were accepted.

On a day-to-day basis, scientists seem to keep going because, gloom and difficulty notwithstanding, they find science fun. Partly, suggests Rubbia, it is a matter of attitude. It may be painful at times, but 'It is sort of a game. Any fundamental advances in our field are made by looking at it with the smile of a child who plays a game.' Peter Mitchell talks in a similar vein: 'whether you turn out to be wrong, or right, everybody wins. This is something else I think that even quite a number of scientists don't seem to appreciate, that being wrong in science is often much more fun than being right, because the next day you wake up with a new horizon . . . It does make a personal impact, of course, but this is something you need to bear because there are more important things than the particular dent in your own personality.' This sense of a serious 'playfulness' crops up again and again. The palaeontologist Elwyn Simons chose to be a scientist because he wanted 'a career doing something that always seems like play'. He continues: 'It's fun to find fossils because you never know what you're going to find and there's always a chance that you'll find something quite unusual, and that kind of excitement makes it sort of like a treasure hunt.' Another field scientist, Jared Diamond, who works in New Guinea, is equally enthusiastic. 'I loved it. Emotionally I'm half New Guinean . . . My identity is very connected with that work . . . it is so incredibly vivid out there. I really love the jungle, which is beautiful, and the birds are fascinating. Tropical biology is very complicated but I learned how to deal with it.'

Of course, it's not only field workers who find science enjoyable. 'That's the trouble,' says the clinician and molecular biologist David Weatherall, 'everything's so interesting . . . in our particular field, although practical things are always on one's mind, there's a tremendous temptation to shoot off into problems of evolution, population and genetics . . . [so you] become a kind of gorgeous amateur at everything.' John Cairns, another molecular biologist, drifted into bacteriology, initially, because he loved 'the smell of bacteria growing on plates. It is a rather nice smell. When you come into the laboratory in the morning there's this homely smell greeting you.' The applied mathematician Sir James Lighthill has spent much of his career working on fluids: 'I have a sort of general pleasurable feel about fluids . . . And my hobby is swimming; I have a great deal of interest in the ocean—ocean waves, ocean currents, ocean tides—and so I enjoy observing all that when I swim. And then I have a fellow feeling for the swimming animals and I've written papers about almost all varieties of swimming fishes and invertebrates'.

A strong sense of an aesthetic pleasure also runs through scientists' accounts of their work. James Black loves chemistry because 'in an imaginative sense, [its] entirely open ended, and entirely pictorial . . . I daydream like mad . . . it's simply something which is a rich food for the imagination. You can have all these structures in your head, turning and tumbling and moving . . . They're just intrinsically beautiful.' Cairns is a committed experimentalist. 'For me the fun of an experiment is like the fun of doing a lot of washing up. You get everything ready and you make damn sure that you can't go wrong. I like to think it's something of an art if one pair of hands gets through a huge amount of work in a short period of time simply by organizing it.' Antonio Garcia Bellido uses a similar analogy: 'the pleasure of finding something that explains the phenomena . . . This feeling is similar in many respects to the feeling of a creative artist who has finished a painting or a musical composition. It's a feeling of having been in agreement with somebody, with nature'. Another Nobel chemist, Roald Hoffmann, who is also a poet, is quite explicit: 'Poetry and a lot of science—theory building, the synthesis of molecules—are creation. They're acts of creation that are accomplished with craftsmanship, with an intensity, a concentration, a detachment, an economy of statement. All of these qualities matter'. There is also 'a search for understanding . . . a valuation of complexity and simplicity, of symmetry, and asymmetry. There is an act of communication'.

For many the aesthetic satisfaction is what sustains them. 'I think the bit I enjoy most', says Anne McLaren, 'is analysing data. If one is presented with a pile of raw data, and can turn it into a satisfying story, that's very enjoyable. I think that's what I enjoy most.' For the biotechnologist, Leroy Hood, 'what I've been impressed with in science over my twenty-one years is . . . there is a simple elegant beauty to the underlying principles, yet when you look into the details it's complex, it's bewildering, it's kind of overwhelming, and I think of beauty in the sense of being able to extract the fundamental elegant principles from the bewildering array of confusing details, and I've felt I was good at doing that, I enjoy doing that.'

But scientific staying power is more than achieving an equilibrium between pain and satisfaction. Another notion which echoes through the collection is that scientists are 'driven'. The conversations are shot through with the metaphors of religious experience, physical dependence and sexual pleasure. 'It's extremely difficult', explains Gerald Edelman, 'to say . . . well, I'm going into the lab again tomorrow because maybe I'll have another epiphany. Experience says forget it . . . that isn't what drives me. Curiosity drives me. I believe that there is a group of scientists who are kind of voyeurs. They have the splendid feeling, almost a lustful feeling, of excitement when a secret of nature is revealed . . . and I would certainly place myself in that group.' Carlo Rubbia describes it as follows: 'We're driven by an impulse which is one of curiosity, which is one of the basic instincts that a man has. So we are . . . driven . . . not by success, but by a sort of passion, namely the desire of understanding better, to possess, if you like, a bigger part of the truth.' Nicole le Douarin, a developmental biologist, was first attracted to the field because 'It is fascinating to start with an egg and twenty-one days later you have a chick . . . and you realize, even if you a very ignorant, that the process is very mysterious. This is what fascinates me, and still does. We are still very far from understanding what is going on.'

This 'driven' quality gives scientists an extraordinary degree of commitment.

Anne McLaren describes herself as 'obsessional'. Hoffmann confesses to being 'addicted to it . . . it's very hard to let go . . . I go in the library on Saturday and Sunday and there are 200 journals. When I come back from a trip that's what I do, sometimes even before reading my mail, I want to see what's out there, what's been done. I love it.' For Rubbia, '[science] is not a job nine to five. When you do science you have to do science 24 hours a day. When you are at home you should be thinking about science; when you are going to bed, you should be dreaming about science. It's full immersion you see.' Michael Berridge says much the same: 'Almost all my waking hours are spent thinking about science . . . to be a successful scientist you really have to worry away at things all the time and even hope, when you're asleep, that your subconscious is continuing to worry away at some of the problems . . . It is a completely all-engrossing, full-time occupation.' Rubbia does not need hobbies. Avrion Mitchison distrusts holidays: 'A long rest, a nice summer holiday, that's not a great way of having good ideas. Good ideas come from exercising the mind.' Carl Djerassi was always able to pursue other interests as well, but only because he was so single minded and efficient. This left him, he freely admits, with little time for '"ad hoc-ery". People almost have to make appointments with you; you even have to make appointments with yourself. If someone you love, for example, says, "I would like to see you", the silent—or, rather, the loud—question is, "About what? How long will it take?"'

It is this sense of consuming commitment which James Lovelock sees as the distinctive quality of good scientists. ' [many] people who are nowadays called scientists are not really scientists, any more than advertising copywriters are literary people. They may be able to write beautiful copy, but it's not quite the same thing. They are in a job, a career. A scientist shouldn't be, I think. A scientist is much more like a creative artist, somebody who does it for a vocation. It's the only way of life they want'.

We could go on. But there is a danger that such exercises draw attention away from the essential individuality of science. These interviews were recorded and presented as series of single and separate conversations because each was intended as a personal exploration and not an attempt to reach great and universal truths. Notwithstanding the inevitable technical details—though we have tried to avoid or elucidate these wherever possible—it is the human qualities of science which come over most strongly: its energy and imaginative richness; the sensations of frustration, love, despair and enchantment which hold its practitioners in its thrall. This is the second collection of our conversations with scientists published by Oxford University Press. We called the first *A Passion for Science* and have deliberately kept the reference to passion in the title because it seems, more than ever, to sum up the power and variety of emotions that these dialogues invite you to share.

A.R.
April 1997

IN TWO MINDS

CARL DJERASSI
*was born in 1923 and is Professor of Chemistry at
Stanford University, California.*

CHAPTER 2

Blemished heroes

❧

Carl Djerassi
Chemist

IT is, to say the least, unusual to find a distinguished scientist who is also successful in the arts. Carl Djerassi, having practically invented the birth control pill, and built up such an art collection that he could afford to donate dozens of works by Paul Klee to the San Francisco Museum of Modern Art, also has a growing reputation as a poet and novelist to add to his list of achievements. It puts him in the unusual position of being able to consider the creative process from the point of view of both the scientist and the artist, and to act as a mediator between the arcane culture of scientific research and life outside. His first novel, *Cantor's Dilemma*, which is set in the high-pressure world of cancer research, attempts to show science as it really is. The public mask of the formal paper is replaced by a very different picture of the scientific process: one coloured by the personal chemistry of temperament, motive and ambition, and darkened by the greyer tones of short cuts, competition and even fraud.

Born in Vienna in 1923, Djerassi trained in the United States, where he has been mainly based ever since. He now divides his life between London, where he spends every summer writing full-time, and California, where he still teaches in the Chemistry Department at Stanford University and has also founded an artists' colony which provides studio space and residencies for up to seventy artists a year. It sounds like the existence of a modern Renaissance man. I wondered whether this was something he had set out deliberately to achieve.

No, but it's probably my character. I'm usually stimulated by doing several very different things. Not exactly concurrently, but perhaps during the same day. This is why even in my scientific work I usually work on several very different projects at the same time. But there's another thing. As you know, scientific research most of the time is just a series of failures. But something every once in a while, of course, works. So if you work on several topics at once, the likelihood is that even though you have some disasters and feel depressed, there is also something you feel good about. I think that's probably another part of the reason why I've been active in various things.

'But that's unusual, because I think most scientists tend to focus very tightly on a particular problem. You seem to have adopted a very different strategy.'

Yes, but you know, I'm a very-well-organized person . . . and that's both good and bad. It's good in the sense that I use my time very efficiently and therefore can do, perhaps, more things per unit of time. It's bad because if you're very well organized you have very little time for 'ad hoc-ery'. People almost have to make appointments with you; you even have to make appointments with yourself. If someone whom you love, for example, says 'I would like to see you', the silent—or, rather, the loud—question is, 'about what? how long will it take?', because it's interrupting your self-imposed schedule. So that is certainly one of the obvious disadvantages, and it took me some time to really learn that.

'Were you always very organized?'

Yes. Except maybe during my childhood. But I was already very well organized by the time I got to college. I was probably the only graduate student at the University of Wisconsin who did not work in the lab at night. I refused to do that, because I wanted to do other things, especially since I was already married at that time. Therefore I did as much work from eight to five, or nine to five, or whatever it was, than others did between, perhaps, ten and eight, or ten and nine at night.

'For you, then, it hasn't been a problem but a pleasure to be active in all these areas. One hasn't interfered with the other?'

No, except as I implied earlier, I think you do pay a price.

'What price?'

Well the fact that I was married three times probably can in part be assigned to this.

'So is the implication that you were so highly organized that you actually had no time for your wife, or your earlier wives?'

It's not so much that I didn't have time for them, but that they didn't speak the same language as I did. They were not scientists, and there was no question during that time that the bulk of my life dealt with very scientific affairs. The theatre, the opera, the music, the art collecting, did all go on but there was much less of it, and I did not write any fiction or any poetry until my second marriage was over. I think that could, perhaps, have made a difference.

'What made you, then, move into the arts?'

Well, 'arts' is a broad word. When it comes to art in the context of art appreciation and music appreciation, that has been with me, probably since my early adulthood. But when you talk about art in the context of actively doing it myself, of writing fiction or poetry, then, literally, I didn't write a word of poetry or prose until my sixtieth birthday.

'So it came very late?'

Oh yes.

'Why did it come then? What made it happen?'

Well, there was an acute reason. It was a traumatic period in my emotional life because I was very much in love with someone who left me. Now, there's a good end to this story because after a year we got together again and we're now married. But during that year I felt very violated. Diane Middelbrook left me for another man. It was unannounced, and I really felt self-pitying and revengeful—all that sort of thing. And for some reason this manifested itself in a very verbal way, in a confessional form of poetry. Now the person I was and am in love with is a professor of literature and poetry herself, so you might say it was a typical macho way of 'revenging' myself on her turf rather than on my own.

'To what extent, though, was your ability to change partly due to the fact that you were financially and professionally secure?'

Well, that had a great deal to do with it. In fact you've probably put your finger on it. It was very important that I didn't have to support myself through my writing and had established a reputation in science that was unlikely to change very much. It gave me the chance to do something totally different. In fact, in terms of intellectual life I can think of nothing that is more different from science than fiction. Fiction is almost the antithesis of science. Scientists never have the luxury of being able to say 'ah, but I made it up'. That's *verboten*. So fiction is a wonderful luxury. You can even brag about it. When you say 'I made it up' people congratulate you! Scientists would kick you out; they would say you are finished.

'Do you, then, see the creative process in arts and science as completely different?'

In some respects I see it as different. First of all there's the manner in which we write about it. In science all we're interested in doing—I'm talking about my own science, chemistry, physics, the hard sciences—is transmitting information. That is all there is to it. You're not supposed to embellish it at all. The moment you embellish it your editors tell you that you're verbose; you must cut it out, be short and succinct.

Your readers aren't interested in anything other than information. They usually read your paper only once, unless they want to repeat some experimental recipe, and then they file it in their mind if it's useful. If it is not useful for them, it becomes mental garbage and gets thrown out. You can't afford to keep it all in your mind. So scientific papers are a very ephemeral form of literature. People very rarely re-read them. But in poetry, on the other hand, the ultimate compliment is to have someone re-read your poems, to remember that book, to remember some metaphor, some nuance which would count for nothing in science.

'That's a major difference in the end product, as it were, but is there any similarity in the actual creative act between art and science?'

Yes, the fact that in both you're doing what hasn't been done before. You flatter yourself that you are the first to open up some terrain, or see something in a new way. Most of the things that fictional artists make up have already been around to a large extent. In one way or another the number of plots can be reduced to very few. So you have to find a new perspective. And that new insight into something is a very exciting but very short-term feeling. It's almost, you know, like sexual pleasure, you have a real high and then it's gone.

'And you get that pleasure in science too?'

Oh, absolutely! I'm absolutely convinced that the pleasure of a real scientific insight—it doesn't have to be a great discovery—is like an orgasm. But it's short. And after a while, the more important it is, the more obvious it becomes in the end. People ask why did you not think of it before? And then—and this is the ultimate compliment in science—they just accept it and forget whose insight it was. It becomes common knowledge.

'I still have no idea, though, what it means to be creative in chemistry.'

Well, it is not different to be creative in chemistry than in any other discipline but, even so, it depends very much on your sub-discipline of chemistry. You have to some extent made a very complicated comment. The real division, and it was certainly true in my own life, is between a synthetic chemist—a chemist who synthesizes something whose structure is known or has been seen—and a chemist who elucidates structures. Someone, in other words, who isolates material from natural sources and then establishes its chemical structure.

I initially started out as a synthetic chemist, and it was during that time that we invented the birth control pill. And here I think the analogy is a very simple one. Synthetic chemists are both architects and builders. We first design a building on paper, and then become the builders who actually construct it. We know all along what we are going to do from having a picture of the house that we want to build, to—unless you're a very unsuccessful builder—eventually constructing it. By contrast, an analytical chemist who is interested in establishing the structure of an unknown compound is in a much more difficult position—one that in the end interested me intellectually much more. Because here at the outset you know nothing about what you're working on. It's like being in a pitch-dark room. First, chemists started feeling around to get an idea of what's in there, what the materials are from which it is made. Later they started to acquire a sort of a candle or penlight and eventually had powerful flashlights. These were the various spectroscopic methods—ultraviolet, infrared, and NMR spectroscopy, mass spectroscopy and the like—which are the chemist's 'flashlights'. Gradually you get to know more and more about the room and eventually you put all the information together, right down to what the colours are and everything else about it. That is intellectually very, very exciting.

It's not the same nowadays, however. It's all changed dramatically through the development of one discovery, one sub-discipline of chemistry, which is particularly highly developed in Great Britain, and that is X-ray crystallography. It's the equivalent of having the little flashlight taken away and suddenly being given a flash camera. You take one quick picture and immediately know everything that's in the room. Thus intellectually the actual structure elucidation process is not very exciting any more, but in terms of information it gives you everything you need. So then the question is, what do you do with that knowledge? Which, translated into chemistry, means what is the physical or biological utility of this structure? How and why does the organism synthesize this particular compound? What can you use it for? What are the advantages, the disadvantages, the dangers? And it's in this area—the practical applications of your chemistry—that the real challenges now lie.

'But I want to pursue this idea of the creative process. When you set out to synthesize the birth control pill, how difficult a problem was it? What was your particular cleverness? Was it, to use your metaphor, the actual design of the building or compound, or its construction?'

Well, in some respects it was both these, but it was something else as well, which was much more important. Let me use an analogy which is a musical one. Say you decide you want us to design a concert hall. We can draw all the plans; tell you how large it will be, what materials are going to be used, and so on, but in the end we can only guess at what we think will be the best design for the ideal acoustics. We will not know whether it works until you play the first concert. Well it was the same with the chemical structure of what became the birth control pill. Endocrinologists had to do the final biological test to see whether what we had predicted did, in fact, happen. And it was the ability to make the right predictions that I think was the important thing. We really didn't know whether we were right until finally that experiment was done.

'What made you choose that problem then?'

To be quite honest, at the time we chose it, the problem was not really directly connected with birth control. What we wanted to do was to synthesize an orally effective progestational hormone. Progesterone is one of the two female sex hormones, and the one that acts as Nature's contraceptive. A woman does not ovulate and become pregnant again during pregnancy because during that time, and only during that time, does she continuously secrete progesterone which inhibits further ovulation. But at the time (1950–1951) we began working on this problem, progesterone was not used for birth control, but to treat menstrual disorders and infertility. Also, and most importantly for us, it was thought to be a potential treatment for certain forms of cancer, particularly cervical cancer. As progesterone itself was only biologically active if given by injection, it meant that large volumes had to be injected locally into the cervix, which was very painful. So we wanted to develop—to invent, really—a synthetic progestational hormone which would still work if it was given by mouth. When we had done that and knew that it really worked, the next jump was, well, what else could you use it for? And the answer to that, of course, turned out to be its widest application: birth control.

'In your novel, *Cantor's Dilemma* you give a picture of science which is very competitive. The image is one of a scientist driven by the desire for personal success and not terribly interested in the science itself. First of all, do you think that's an accurate description of the novel?'

It is not. The first part is quite correct, about the drive and so on, but I don't think it correct to say that its protagonist, Professor Cantor, was not particularly interested in science itself. This may be a reflection of the way I wrote the novel. Instead of writing about science itself, which many people such as science journalists do very well, I wanted to do what very few authors do, and that was write about the behaviour and culture of scientists. And because that's what I wanted to emphasize, I underplayed the actual science that Cantor and his student were doing. But I'll tell you this: the drive of the scientist—or, rather, of the super-scientist because I'm not talking about all scientists here, only about the élite—that drive to succeed means that the élite is necessarily blemished. And I put myself in this category—I'm not saying this just about other people. But we should not be ashamed of this because the urge to compete is also the fuel, I think, that really leads to many of these achievements.

'You mean the ambition is also a blemish?'

Absolutely, in the context in which you meant it, namely, why not do science for science's sake? I do not believe that there are very many scientists who, if they were very honest, would say they only do it for its own sake. Because if that really were the case, you could say 'all right, but publish anonymously'. Very few would; not for the salaries that they're getting. That is actually the main topic of my second novel, *The Bourbaki Gambit*.

'You also give an impression in *Cantor's Dilemma* that great ideas, or important ideas, come almost by the "eureka" effect; that you're sitting quietly and suddenly you see it. Do you think that's true?'

Great discoveries, real intellectual conceptual breakthroughs, do have that moment of clarity, yes. Now, most of scientific advances do not occur that way, but remember that I'm talking about super-scientists, and really sensational discoveries. The DNA structure, for example. There must have been a 'eureka' moment in the life of Watson and Crick.

'I think I take a slightly contrary view because there is the story of Newton who, when asked how he arrived at the theory of gravity, said "By thinking about it continually." And so I'm always a little worried that the cry of "eureka" conceals the enormous amount of work that has to precede it.'

I don't argue with you at all on that, I really don't.

'Another feature in the novel is that you give a feeling of the greyness of scientific observation.'

Well, that really has become a preoccupation in my present teaching. I really believe that the colour we use least in our metaphors is grey. In our political metaphors, we talk about the 'greens', we talk about the 'brown shirts', we talk about the 'blacks', the 'reds', we rarely talk about grey, except, perhaps, when referring to old people. Yet grey in my opinion is *the* politically realistic colour of this century. All really major problems that we have to face are grey problems, none of their solutions are black and white solutions, they are all grey. We do not want to listen to grey questions. We do not want to hear grey answers. And that is what I want to talk about in a way.

'Does your novel really reflect your own experience of science?'

It reflects it in a number of ways. First, of course, the drive for the Nobel prize. I've been a foreign member of the Royal Swedish Academy of Sciences for many years, and the only privilege I have as a foreign member is to be able to nominate people for the Nobel prize every year. One cannot vote for them as a foreign member, but one can nominate them. So I do know what is involved, and how people really count on it and use it. So in that context the book is realistic. It is basically about noble science and Nobel lust.

Another message in the book is the importance of the mentor–disciple relationship, and how the disciple really learns almost everything from the scientific behaviour of his mentor. And I purposely say 'his' right now because it's still very male oriented in that context. The learning comes about through a form of osmosis, which also works both ways. The mentor who has a star student really has the feeling that he has a son. Hopefully the day is coming when we have more mothers and daughters. There is an umbilical cord in this situation that is stronger than in almost any other situation where you have mentors and disciples; an umbilical cord that stretches but hardly ever gets cut even when the mentor dies.

I think the only event in my novel that I have not experienced personally is being involved directly in a question of possible fraud. That has not happened to me, thank God, in my own work. But I would say, 'there but for the grace of God go I', because the sequence of events described in the book and the way things have happened on a number of occasions, even quite recently, could happen to almost anyone and we'd better be aware of that. I think in the United States, particularly, the question of fraud—and I'm now talking about real fraud rather than the grey area of massaging results or something like that—is, in my opinion, very rare, a real aberration. But I think we should realize that we do take shortcuts, that we misbehave in this regard once in a while, though people have assumed that scientists are totally unblemished in this context. That is absurd—no group is perfect.

'How was your novel received by your colleagues?'

You hear me humming and hawing, because the reaction was mixed. I would say first of all that quite a number of my truly intimate colleagues probably never read it, which shows that chemists, in particular, just don't read fiction. Nevertheless a lot of scientists have read it and there were some people who really felt that I was washing dirty lab coats in public. I really felt that a lab coat is only a lab coat when it gets dirty. In that case there's nothing wrong with doing the laundry in public and people really should know about the dirt. I didn't feel that I really maligned our discipline or anything like that, so I consider myself not guilty of that charge. It's also interesting that *Cantor's Dilemma* was reviewed very widely in the States both in the scientific literature and general press. On the whole the reviews have been very complimentary, but a few scientific ones, in fact right here in England, were much more pricklish. But that's fine, because I feel that I've touched a raw nerve and in parts I wanted to do that.

'You have many honours. Do you think the honours system is good for science?'

Yes and no. The system goes some way towards satisfying an unsatisfiable drive. I have yet to meet a scientist who thinks he's gotten enough honours. Even those people who have won a Nobel prize would like to win a second Nobel prize and maybe some of them would like to have a third one, though no one has gotten three of them so far. I would say in some respects, yes, honours are a good thing. But in other respects, no, because we overemphasize the importance of honours. Many people who have not gotten them deserve them just as much, but are not really treated as well as they should be. Someone, I forget who, said quite correctly: 'The Nobel prize is very good for science, but terrible for scientists.'

'Winning a Nobel prize, of course, is a very potent prospect. Have you been disappointed that you haven't had one?'

Rather than as a 'yes' or 'no', I prefer to answer that in a slightly more complicated way. To me the Nobel prize, as it is to everyone else, is the paradigm of the ultimate scientific honour, and most scientists who have made some very important discoveries or have worked at the cutting edge of science probably dream at some stage or other that it may be very nice to have won a Nobel prize. Now there is no question that I belong to that group of dreamers. I would be a liar if I'd said 'no'. On the other hand, I also know that there are many more people who deserve a Nobel prize than could possibly get it and I guess I'm probably in that group. It is not really one of the big tragedies of my life, because I have won a number of other awards. But again, like most other scientists, I would like to get just one more, it's a never-ending thing. On the other hand, I am now moving into literature, so I would be just as pleased to get a literature award here or there. In fact it would probably please me more than a considerable number of larger scientific awards, excluding the Nobel prize of course.

ROALD HOFFMANN
was born in 1937 and is John A. Newman Professor of
Physical Science at Cornell University, New York.

CHAPTER 3

Tapeworm quadrilles

$\underleftrightarrow{}$

Roald Hoffmann
Theoretical chemist

THE notion of reductionism is essential to modern science. It is the belief that the properties of everything, including the brain, can ultimately be explained in terms of physics.

It is surprising, therefore, to find that Roald Hoffmann describes himself as an anti-reductionist. He and his collaborator, Robert B. Woodward, did after all win the Nobel prize for applying quantum mechanics to chemistry. It is also surprising, though perhaps it shouldn't be, that a distinguished chemist is also an established poet. Hoffmann's two volumes of verse, both published in the past few years, have received considerable acclaim.

I set out to ask him about both his chemistry and his poetry. He is a theoretical chemist rather than one who cooks up things in the lab, and I wanted to know what that actually means. I also wanted to know how doing chemistry compares with writing poetry, do the two activities have anything in common? First, however, the beginnings of both. Hoffmann was born in Poland in 1937. How did he get into his chosen fields?

I could be a good piece of propaganda for the immigrant ethic of what the United States does for people. 1937 was not a particularly good time to be born Jewish in Poland. The family was very happy, but around us there were gathering storms, and we had a very difficult time during the war. My father and many others were killed by the Nazis. Eventually we got out of Poland in '46 and came to the United States in '49, after a number of years as refugees in Europe. So, I'm one of the last generation of Hitler's gifts to America! I got into one of these marvellous science-oriented high schools in New York City, mine was Stuyvesant High School, another one is Bronx High School for Science; many scientists have gone to these élite, yet state, schools. I wanted to do mathematics, but just one year in high school showed me there were people who were better at mathematics than I was, and at the end of high school I guess I was to some extent sure of science, but certainly not committed to chemistry at all. I had done my share of chemistry experiments at home when I was a teenager, but I can't say there was an abiding urge to become a chemist. Even when I was at Harvard in my first year of graduate school I sat in on courses in astronomy (planetary atmospheres, I remember), and a course of science policy. And I applied to a summer programme in archaeology to do some excavations in Turkey. I think all of this is evidence that I wasn't sure what I wanted to do.

'You were already a chemist at graduate school?'

Yes, I was at graduate school in chemistry. I had made that decision but I wasn't that sure that that's what I really wanted to do. I think it's just by a hair that I became a chemist, I could have become, well, something else.

'Yet you speak in your writing about your love for chemistry and how beautiful it is.'

Yes, I love it. I like the subject, its richness, its position in between, its compromise between simplicity and complexity. Chemistry is positioned between the simple world of atoms and the complex nature of biological molecules and real materials. This is something I like. Because I think in science most scientists are pulled to the simple, they simplify the world around them. They set up a box of those problems which admit of a unique solution, of a resolution, and then they do very well at solving those problems. And then there is complex reality, and a coming to terms with the world as it is around us, which is awfully complicated. I like that, I think chemistry is nicely positioned in the middle. You have to talk about molecules that are simple, and you have to talk about haemoglobin, something that looks messy, like a clump of pasta that congealed from primordial soup, like a tapeworm quadrille. And you can't run a body with hydrogen and 92 elements, you need molecules, you need thousands, millions of molecules, you need complexity. And then, on the other hand, there's the human mind which always tries to reduce to simplicity.

'But you're a theoretician?'

Yes.

'But you did do experiments?'

Not many. The first two papers I published are experimental papers, one on thermochemistry of cement, another one on some radiochemistry. They're essentially juvenilia, they come from things I did when I was at college and summer jobs. I'm a theoretician but I'm very close to experiment. I take my inspiration from experiment, I try to explain things that other people have found, and then I build general theoretical frameworks, but I'm always very close to experiment.

'I need to understand the relationship between the theoretician and the experimentalist. Is there a tension there?'

I think theory and experiment are always in contention, there is a love and hate relationship between the two. The experimentalists say that theorists build castles in the sky, they don't pay any attention to reality; the theorists say the experiments aren't the right kind of experiments to test their theories. Then we also have a framework of reductionism which values theory in excess, makes us believe that understanding is theoretical. And this occurs in every field of human endeavour, not just in science. We have lines drawn between people who build fancy economic models, very mathematical, and people who deal with the realities of the stock market. We have critics and writers in English departments. But the truth is that they desperately need each other, which is why they are in contention, of course. Theory and experiment in chemistry are in such a love–hate relationship too. Most theorists are in a business of rationalizing, that's the most kind way of saying it, things that other people have found experimentally. Now, I like to play the game a little bit in another way. I think that a theoretician can get out of the situation where he or she is just helping an experimentalist by making slightly unreasonable predictions. The operative word is 'slightly' here. Very unreasonable predictions that take ten years' work on the part of somebody will never be tested. 'Slightly' is defined probably on a time scale of a graduate student's existence in graduate school. Anything that takes one to three years to do, a research director is likely to devote the effort. I think sometimes that I'm very good at making such slightly unreasonable predictions so that people test them. I love that interaction, with experiment; this is very close to the heart of what I do and what motivates me.

'Are these theories of yours general, like a physical theory, a theory of everything, or is it that you take many different chemical reactions and provide a theory for each of them?'

I take many different chemical reactions. I am a firm believer of moving from specifics to generalities, not of building general theories. First of all, the work that I do is almost amoeboid-like in character; I'm reaching out to various pieces of the chemical world. One day it's copper gallium compounds, another day it's a molecule that's shaped like a triple helix built up from tetrahedra—there is one that's standing in a model on my desk here. Another time it's what happens to aluminum alkyls, compounds of aluminum falling apart on a metal surface, as they fall apart. I take these individual problems and they're like pseudopodia going out in a different direction. Underneath I have a theoretical framework which happens to be called molecular orbital theory, of how electrons move in molecules. I have a way of looking at my molecules. I have also a conviction that everything in the world is connected to everything else, and if I send out enough of these pseudopodia they'll merge into something which will be an understanding of all of it. By going off in different directions I guarantee for myself that I am not locked in on one set of compounds, that I am forced to see the relationships between different ones. I think the beauty is in the complexity of nature. So I come back to something I said before, that beauty is in the reality of what's out there, residing at the tense edge where simplicity and complexity contend.

'It's a slight paradox because I have a feeling, both from what you've said and from your writing, that you're somewhat anti-reductionist.'

Yes.

'Yet your very work by applying quantum mechanics to chemistry is what any naïve person like myself would call pure reductionism.'

That's right, there is a paradox in this. I am a theorist, I'm in a business of providing explanations, my tool is quantum mechanics, a method of physics applied to chemistry, but the way I use quantum mechanics would drive most physicists crazy.

'Are you against reductionism in general?'

Oh yes. I think reductionism is unrealistic, it's just an ideology that science has bought. I think understanding comes in two types, horizontal and vertical. Vertical understanding is reductionist understanding, or analysis. Horizontal understanding is the understanding of a concept in a field in terms of concepts of equal complexity, of equal categories. Let me give an example of a *reductio ad absurdum* example which you need not accept. If someone sends you a poem, let's say, the phrase from T. S. Eliot's *Murder in the Cathedral*, 'The last temptation is the greatest treason; to do the right deed for the wrong reason.' Let's say you get that in the mail, an unsigned poem. To ask what is the sequence of firing of neurones in somebody's mind when they read that poem, and the biochemical actions behind it, and then the wondrous chemistry and physics behind that; knowing that will get you a lot of Nobel prizes. But it has nothing to do with understanding that poem—when it was written by Eliot, or read by the reader, or the person who sent it. That understanding is on a level of the English language and the psychology of the moment. That's horizontal understanding to me. What I'd like to say is that even within two fields as close to each other as chemistry and physics, embedded at the heart of science, that there are concepts in chemistry which are similarly not reducible to those in physics, or if they are so reduced they lose, just like the poem, everything that's interesting about them.

'That comes as quite a shock to me, but to turn to your analogy with poetry, you are an established poet, how did you become one?'

I didn't try to write one until I was about forty or so. I always thought it's an interesting way to try to understand the universe around us. And I then tried writing. I wrote in a vacuum for myself, I made a mistake, I should have taken a course. I was too ashamed to take a course because when you're forty and an established professor, you don't do that. Or so I thought. So I wrote poems. I sent them out and they came back with rejection slips. These are not like referees' reports on a scientific paper: nothing comes back, just a piece of paper saying 'no', and you can paper your bathroom with such pieces of paper because the acceptance rate of poems is much lower than it is for science. Scientists sometimes complain, but in a major American journal of chemistry, the best one in the world, the *Journal of the American Chemical Society*, two-thirds of articles get accepted. In a routine poetry journal, far from being the best, five per cent of poems that get submitted get accepted, and in the best journals one-half of one per cent. So poetry is a much more difficult business to break into. Or else these statistics are telling us that more people are writing poems than are willing to buy poetry journals! Anyway, I began to write poems. I got all rejections. It took me seven years to get a poem published from when I first wrote one. Eventually I met a group of people here at Cornell, among them a wonderful American poet, Archie Ammons, and we began to talk and read for each other, and in subtle ways I got some criticism and feedback. Since then I've published a number of poems and two books.

'Is it an important part of your activity?'

It is. It was a way for me to break out of science in a way, even though there are connections between the poetry and science, but it gave me something to do, it gave me a certain concentration.

'But why did you need this, I don't understand the need to break out of science.'

I don't understand it either, Lewis. Something happens to men around forty perhaps. They do worse things than write poetry. Something, I think, was in me all along, that this was, as I said, a valued way of summarizing the essences of this world. Now, putting it that way sounds almost like doing science, but not quite. I think there are similarities; there are perhaps differences. I just felt the need for writing poetry.

'What are the similarities and differences between doing poetry and doing your sort of chemistry?'

I think poetry and a lot of science—theory building, the synthesis of molecules—are creation. They're acts of creation that are accomplished with craftsmanship, with an intensity, a concentration, a detachment, an economy of statement. All of these qualities matter in science and art. There is an aesthetic at work, there is a search for understanding. There is valuation of complexity and simplicity, of symmetry, and asymmetry. There is an act of communication, of speaking to others. Those are the things that I see are similar. What is it that's different? One thing that's different is, and here I borrow a line from Gunther Stent, that science is infinitely paraphraseable and art is not.

'What do you mean by that?'

Well, I will explain. If you discover how to make a new drug, an immunosuppressant, and you write up the synthetic procedure in English, maybe translate it into Japanese, a factory can be built for making that drug around the world. But you take a poem and you translate it, at best it's another poem.

'But do you think that your poems really offer explanations the way your theories in chemistry do?'

No, the poems where I've tried consciously to offer explanations are, by and large, not successful. Poetry is at its best when it tackles problems which have many different resolutions rather than one solution. That's another difference that people have pointed to between the arts and the sciences. There are so many different kinds of poems! Some arise from the sound of a word, some arise for me in fact from science, from seeing a metaphor in the natural world for something in the emotions. Let me give you an example. The title of my first book is *The Metamict State*. This is a rather unusual state of matter—not many scientists know what the word means. The metamict state refers to crystals of radioactive minerals. If you were to grow a crystal of a uranium salt, it would form, for some salts, a beautiful colourless crystal. Then with time the atoms would start decaying, the nuclei would fall apart. The uranium atoms under the influence of all that loosed energy would move off their lattice sites, they would knock into other ones and slowly the order of that pristine, clear crystal would be destroyed. The crystal grows yellow and amorphous. The enemy is within.

Now, that's a metaphor; even as I say it, it forms a poem. I listen closely to scientists. It seems to be given to us to attend dull seminars, and when this happens to me (does it to you?) I defocus from the substance of what is being said and I focus on how it is being said. The language of science is incredibly interesting; it's a natural language under stress. The language is put under stress to explain things that are difficult to explain, that perhaps are explainable in terms of mathematical equations and structures. In this language simple words like power, energy, force, stable, unstable, acquire a host of alternative meanings. That's partly what poetry is about, about ambiguity, about alternative meanings.

'Well, that's just the point I wanted to pick up because you've contrasted the perfection of science with the imperfection of scientists. What did you mean by that?'

Well, science is a wonderful system, a Western invention for getting reliable knowledge. The system, like a number of other social inventions, works remarkably well with imperfect people, harnessing their normal, natural psychological forces, which may be for advancement, for recognition, for praise. A tension results. I find it interesting to step back and look at that tension. Take the scientific paper. I'm not the only one who's been concerned about it—Peter Medawar has spoken in the same vein. Here is the journal report, a product of 200 years of ritual evolution, intended, supposedly, to present the facts and nothing but the facts dispassionately, without emotional involvement, without history, without motivation, just the facts. Well, underneath there's a human-being screaming that I'm right and you're wrong. That endows that scientific article with an incredible amount of tension. Now, you have to be a scientist sometimes to know where the tension is; sometimes it's more important to know who has been omitted among the first ten footnotes rather than who is included—very much like in Russian communist days taking stock of the line up of people standing on top of Lenin's tomb. When you then create a way for things to come out anonymously, as you do in our wonderful refereeing process, you are setting loose some of those repressed human forces which are underneath, and do they come out! I have somewhere a collection of referees' comments on my papers which I sometimes show to my graduate students with some trepidation. They, the students, are still in a romantic phase, they don't really think there are people out there who could possibly say such nasty things about me.

'You've had really nasty referees' reports?'

Oh sure, sure.

Well, all I have to say is, how do you see your own future? Will you devote more and more time to poetry and to writing?'

I'm stretched pretty thin as it is in terms of my time because I keep on doing the science, I'm addicted to it, it's very hard to let go of it. Sometimes I wonder what will happen. I go in the library on Saturday and Sunday and there are 200 journals. When I come back from a trip that's what I do, sometimes even before reading my mail, I want to see what's out there, what's been done, I love it. It is a kind of addiction. For instance, articles that I've written in the last few years have been cited by people thousands of times. Now, it's wonderful when somebody you don't know at all cites an article, it's much more interesting than when somebody whom you know cites it. That kind of interaction is very addictive, it's very hard to stop that maelstrom of activity. I wish I had the strength sometimes to stop it, I don't. I would like to do more of the writing. I think I'll do more of the science too. Both!

JARED DIAMOND
was born in 1937 and is Professor of Physiology at the
University of California, Los Angeles.

CHAPTER 4

Worlds apart

⟨⟩

Jared Diamond
Physiologist and evolutionary biologist

THERE is often little love lost between the exponents of the so-called hard sciences—physics, chemistry, molecular biology and the like—and their colleagues in what they think of as the softer disciplines, such as ecology. The experimental camp considers the field sciences to be little more than stamp collecting, while the massed ranks of botanists and bird-watchers are equally uneasy about limitations of the reductionist approach. The two philosophies rarely occupy the same building, let alone the same department. So, to find an individual who subscribes to both is remarkable.

One Jared Diamond is a Professor of Physiology with a distinguished record of research into the behaviour of cell membranes. The other Jared Diamond is an equally distinguished ornithologist, ecologist and explorer who periodically abandons the laboratory bench for the jungles of New Guinea. How did he acquire this divided self?

———

It began very early, through a convoluted background. My father is a physician, and so when I was growing up and people asked me what was I going to do, I always said, 'I'm going to be a doctor like my father.' But I had a schoolteacher when I was eight years old who was a bird-watcher, so I became a bird-watcher when I was young. When I got my Ph.D. in physiology in Cambridge and came back to the United States, I had a problem with my image as a scientist. As a child I'd been interested in lots of things. My mother was, and is, a musician, so I was a pianist. I love music. My mother was also an accomplished linguist, and in Europe I loved picking up languages by ear. I nearly dropped out of physiology at the end of my first year in Cambridge to become a linguist. I also loved writing. There were lots of things that I was interested in so that the idea of confining myself to physiology or any single scientific discipline for the rest of my life sounded awful, and yet this is what you're expected to do. But when I came back to Boston as a postdoc from Cambridge, I was lucky that I had a fellowship which gave me a good deal of freedom, and did not tie me to a particular laboratory. At the end of my first year as a postdoc I simply told my laboratory director that I was going to go to the Amazon to study birds. For four consecutive summers I went, first, to the Peruvian Amazon, and then to New Guinea, studying birds, without any special training, just observing them.

'Is that really so different to physiology?'

Totally different. To begin with, it's not a laboratory science, it's a field science. It's not an experimental science, it's an observational science with a historical component. I'd had formal training in physiology, I never had any formal training in ecology and evolution.

'But that first expedition that you went on to New Guinea, what were you looking for? And why New Guinea?'

Why New Guinea? My friend John Terborgh and I decided we wanted to go to the Tropics because Alfred Russel Wallace, and Darwin, had shown that the tropics are the regions of greatest biological diversity, it was the place for a biologist. Where in the Tropics to go? Well, the wildest place in the world in the early 1960s was New Guinea. There were still tens of thousands of uncontacted Stone Age people there, and my father had been the mentor of Carlton Gajdusek, the brilliant virologist at the National Institutes of Health, who carried out medical exploration in New Guinea and had made first contact with tens of thousands of New Guineans. Dad introduced me to Carlton, and Carlton told me how to operate in New Guinea. So, I had some clues what to do there. John and I went out to New Guinea extremely naïve. In order to raise grant funds, we basically cooked up a project. John realized from the literature that tropical birds were supposed to have more of their eggs lost to predators than temperate birds, and also that tropical birds lay fewer eggs at a time than temperate birds, so there was a theory that the small clutch size is related to the risk of predation. John and I then said that we were going to find nests of tropical birds, watch them and see what fraction of them got predated. We actually got some money from, God bless them, the American Philosophical Society. Despite our lack of credentials they gave us $2000.

In the first two weeks we never found a single bird's nest. That project was a dismal failure, and that then left us in New Guinea for three months, interested in bird nests but without a project. So, we watched birds. John became interested in birds feeding in flowering and fruiting trees, and I found myself interested in the altitudinal distribution of birds, and so when I went back the second time, I then proposed to study altitudinal distributions. That then gave me the idea of looking at isolated mountain ranges where altitudinal distributions shift. The purpose of my third trip was to visit the largest unexplored mountain range in Papua New Guinea, a mountain range where, in fact, it was not even known where and how high the highest mountain was. So, on that trip I ended up discovering the highest mountain, climbing it with a great deal of effort, and discovering new birds on that mountain. That's an example of how on each trip, I ended up working on things that had attracted me on previous trips.

'You clearly liked living in the wild in New Guinea?'

I loved it. Emotionally I'm half New Guinean, although nowadays I only spend a month every two years in New Guinea.

'What do you mean by you're emotionally half New Guinean?'

My identity is very connected with that work in New Guinea. It is so incredibly vivid out there. I really love the jungle, which is beautiful, and the birds are fascinating. Tropical biology is very complicated but I learnt how to deal with it. And New Guineans themselves are fascinating people. These are people who were in the Stone Age a couple of decades ago. Many of the people that I've worked with grew up in that Stone Age, and participated in the last inter-tribal wars. This is what the whole world was like until 10 000 years ago, but it's a way of life that's vanishing, and this is the last piece of it still here to see.

'How then did you reconcile in your mind the New Guinea side of you and the hard-nosed molecular physiologist?'

I was interested in both. I was still interested in molecular physiology, and I was also interested in evolution. As I mentioned, the idea of doing one thing did not appeal to me, so I wanted to do a couple of different things. There's also a practical consideration. There are lots of careers in molecular biology and physiology, and there is generous grant support. But there are few careers in evolutionary biology and field work, and there is poor grant support. In that respect, if my goal in life had been to become a New Guinea explorer, in retrospect I should have done what I actually did, that is to get a Ph.D. in physiology, get a secure job, get grants in physiology, and then, on the side, carry out my evolutionary work.

'But did you not feel at any stage torn between these two worlds, or was it not strange going back to the lab each time? They must be very different.'

There are a lot of differences between physiologists and reductionist biologists on the one hand, and field biologists on the other hand. One is that field biologists tend to go into the field because of love of the material itself, separate from the intellectual questions. I love watching birds, and I would study birds in New Guinea even if there was no intellectual payoff to it. It just happens that there is an intellectual payoff, but that's gravy, that's a nice extra. In the case of physiology, the only reason that I would have anything to do with a physiological laboratory would be in order to answer the interesting questions that come out of it. Granted, there are physiologists that love instrumentation and that love experiments, I'm not one of them. What I like is the questions and understanding the organization of the body. I think in general that field biologists love their subject matter, whereas reductionist biologists enter into their field for the intellectual questions.

'You discovered the bowerbird?'

Yes and no. There are eighteen species of bowerbirds of which one was 'lost'.

The 'lost' bowerbird—this is an interesting story in its own right. Bowerbirds and birds of paradise are birds with spectacular plumage confined to New Guinea. Most of them were discovered in the last century, collected by natives because of the spectacular plumage, and then sold to European plume dealers who shipped them to Europe where the plumes were used to decorate the hats of Europeans. The result was that birds of paradise and bowerbirds were described by ornithologists on the basis of plumes and skins that turned up in Paris hat shops without any knowledge where they came from. Eventually, scientific collecting expeditions started going out to New Guinea and actually finding in particular places in New Guinea the birds with the plumes that had showed up in the hat shops. Eventually all these hat shop birds were tracked down except for one of them, the Golden Fronted Bowerbird. Expeditions that went out to the wildest imaginable places in New Guinea didn't find the bird, and by the 1960s the last targeted effort to find the bird was by my predecessor at the American Museum, Tom Gilliard. Tom went out to Batanta Island in 1964 where there were reports of a bird that nested on the ground. He thought, aha! a bowerbird. It was a shattering expedition. Tom got sick, the bird proved not to be the bowerbird but an ordinary pigeon, and Tom died of malaria the following year. People began to wonder whether the bird was extinct, or was in some really remote area.

In 1979 I was invited into Indonesian New Guinea to help the Indonesians set up their National Park system, and I arranged to go into different remote areas of Indonesian New Guinea that were not biologically explored, but that might be important as National Parks. One of these was the largest remaining unexplored mountain range in Indonesian New Guinea, the so-called Foja mountains. They were so remote that there was no human population. The only way to get in was by helicopter, and we went in '81 and were dropped off by helicopter at high elevations. As soon as I entered the forest, literally the first bird that I saw was the Golden Fronted Bowerbird. On the second day I discovered the bird's bower, and on the third day I discovered the mating display of the bowerbird. That was how I discovered the long lost bowerbird.

'The peculiarity of bowerbirds, as I understand it, is they have this little house where they put little ornaments.'

That's correct. Of the eighteen species of bowerbirds, fourteen of them were already known to have bowers. As for the fifteenth, the long lost bowerbird, it was not known but was assumed that it too had a bower. And the discovery of its bower was of particular significance. It was not just that, gee, it's great to discover this lost bird, cheep, cheep, romantic, romantic. The lost bird belonged to a group of five bowerbirds which varied in the fanciness of their plumage. Among the four that were already known, the species in which the male had the fanciest and most beautiful display plumage also had the most ordinary, uninteresting bower. Conversely, the species in which the male had no display plumage whatsoever had the most incredibly fancy bower. With the bower of the fifth species, the Golden Fronted Bowerbird, still unknown, that fifth species was a test. It was a relatively fancy species, and it ought to have a somewhat lousy bower. This was then a test of the hypothesis that in the course of bowerbird evolution the female's attention had been gradually transferred from the male's fancy plumage to the male's fancy bower.

Here was a bird in which the male has a gorgeous 4-inch golden orange crest on its head; when I discovered its bower, it turned out to be a lousy bower by the standards of bowerbirds. By lousy bower, I mean that it was a tower of sticks 6 feet high, built around a circular moss platform a yard in diameter with a parapet and neat little piles of blue, yellow, and green berries on the moss circle, but by the standards of bowerbirds that's lousy. Compare it with the bowerbird population that I discovered in 1983, in which the male has no ornamental plumage whatsoever. It has a good bower—a tower of sticks 9 feet high, the sticks are not just woven together but they're glued together, the whole structure is painted a shiny black, the platform is decorated not just with these lousy blue, yellow and green decorations, but with butterfly wings and 400 acorns and with 300 snail shells and with sticks 6 feet long propped up at an angle of 45 degrees, and with painted black beetle shells. That's a really good bower.

'I think it was Rutherford who somewhat contemptuously said that biologists like yourself are simply stamp collectors. How do you respond to that view? I wonder if some of your physiologist colleagues may feel the same?'

Rutherford exemplifies a wilful and, to me, despicable and destructive ignorance on the part of many scientists about the field sciences and the historical sciences. It's not only Rutherford and other physicists, but many biologists. I would say most biologists, who are experimental laboratory biologists, do not understand the importance of the field sciences and do not appreciate our methods. Their belief is that if you cannot do it in the laboratory by controlled experiments, then the other science is inferior. They believe that observational science, descriptive science, as Rutherford said, is postage stamp collecting. To that I would say several things. Science is a matter of obtaining knowledge by whatever methods work for obtaining that knowledge. The questions of ecology and evolution and historical sciences generally can be answered only by observational comparative methods, not by experiments. That's one thing. And a second thing is that if you want to talk about descriptive non-intellectual science, most of modern molecular biology and experimental science is just that; it is descriptive, often devoid of intellectually interesting questions, like simply describing the ten-thousandth cloned gene. They're in it to clone the gene rather than to ask interesting questions, just as nineteenth-century butterfly collectors were in it to collect the butterfly rather than to ask the interesting questions. The gene cloners are defended today by saying that having cloned the gene, someone, perhaps not them, can then ask interesting questions. Similarly with the butterflies.

The final thing that I would add when I said that Rutherford's attitude and that of many modern biologists is not only ignorant but despicable and destructive is that, as far as society is concerned, the most important questions in science today, the most important biological questions, the ones that have the greatest importance for society, are not questions that are going to be answered by experimental biology. They are instead questions of the biological environment on the one hand, and of human environmentally triggered diseases on the other hand. The thing that really risks making our world not worth living in, or viable for humans fifty years from now, is the progressive collapse of our biological environment. So, field science, ecology, and evolution are, I think, the most important areas of biology for the continued existence of human society. But suppose you want to take an even narrower view and say, 'I'm not concerned about human society fifty years from now, that's too big a question. What we really care about is human health, after all what's important is to cure cancer, and that's what molecular biology will do.' Again the fact is that, in Western industrialized society, most of us will die of diseases of modern life-styles, environmentally triggered diseases. What's important is to prevent these diseases, not to attempt to cure them once we've contracted them, and for that reason too, environmental sciences such as epidemiology are more important for preventive approaches to human health than are interventionist approaches.

'But isn't there a danger, though, that those sciences are essentially soft? You can sort of make nice stories, but it's very hard to know whether you're right or not, or do you think I'm being unfair?'

Yes, you are being unfair. The word soft is used as a pejorative. If by soft you mean sciences that don't lend themselves to the experimental approach, then soft is a neutral descriptor, it should not be a pejorative. Yes, since you cannot do experimental interventions, it's harder to establish what cause and effect are, but there still are ways to establish cause and effect. The most important intellectual contribution to all biology is Darwin's theory of evolution. It's not just a theory: every intelligent person recognizes that modern organisms have evolved from organisms in the past, and the underlying mechanisms are natural selection, sexual selection, drift, and so on. The mechanisms are established but one can still argue about their relative importance. So, yes, there are ways to understand cause and effect in the past, although it's obviously not as easy as pouring things into a test tube and doing the experiment in a day.

'But when you were a field worker, what would you say your skill was?'

As a field worker? Let me put it impersonally. Field workers require several skills, and I think they're all things that come comfortably to me, which is why I enjoy field work. One is being able to detect interesting questions in a mass of complexity. For example, consider the experience of being in a jungle. You're out in a jungle, everything is green around you, the canopy is 120 feet high, you don't see any birds or animals or snakes anywhere. The trees all look similar and you're uncomfortable. What intellectual question is there to answer there? In the case of the jungle some of the interesting questions concern why there are so many species in the jungle compared to in Britain. Have they evolved in the last 100 000 years, or do they have histories of millions of years? These are some of the intellectual questions, but it's a big step to translate your experience of sitting there sweating in this jungle where you can't see any birds, to go from there to answering those big questions. One thing that's required for a field biologist is to see what the questions are.

A second requirement is observational powers and observational skills. In the jungle in New Guinea you hear most birds, you don't see them, and so you have to be good at identifying birds by sound. Since I'm musical it happens that I have a good ear for bird sounds. Just as an example, in 1986 when I went to New Guinea, along with me were the ecologists Paul Ehrlich from Stanford and Richard Southwood from Oxford. One morning I took them out into rainforest for their first experience of observing New Guinea birds in rainforest. We got out there before dawn, and by 7.30 a.m., I had identified fifty-seven bird species for them, but we had not yet seen a single bird and finally at 7.30 we saw our first bird species. All of those fifty-seven birds until then were birds that I had heard and identified by voice. So, another thing required for field work, especially in the tropics where the birds are at the top of the jungle or in dense vegetation, are observational powers, learning to identify by sound, or being able to detect a bird just as a quick bit of motion that's different from the motion of a leaf in the treetops.

Then a third thing is just synthetic ability. To understand the jungle you can't operate narrowly. Again let's contrast it with modern molecular biology. You can make a lot of headway in modern molecular biology by being very good at very little and being massively ignorant about everything else. In order to understand ecological problems, there are lots of things to fit together. You can't understand New Guinea birds without knowing something about botany, or geology. You have to know something about the people because the distributions of birds may be artefacts of what people have been doing for the last 10 000 years. So you have to have a breadth of knowledge and be able to synthesize it.

'Because of your unusual career—an explorer physiologist, an explorer evolutionary physiologist—you are very special in having occupied these two worlds. Why do you think you're such a rare but well-read bird yourself?'

It's hard, it's very hard for a scientist to make a go of it in more than one area. The cards are just so heavily stacked against you. First of all, there is the prejudice that if you work in a second area, you're a dilettante. Secondly, there are the realistic difficulties. How are you going to get grant funding in a second area where you don't yet have any indications of ability? There's also a question of the outlook. In scientific papers we are forced to trim speculation from our papers, we are forced to write them in a dull style and we're prevented from making connections. So we're prevented from working in a field other than our own field.

'Have you had difficulty with your colleagues?'

I have had no difficulty with my colleagues, and for that I am infinitely grateful to my colleagues here at the University of California Los Angeles Medical School. They accepted from the beginning that I would be devoting a part of my time to evolutionary biology. They were willing to tolerate it because my work in physiology was also going well. Conversely, my ecological colleagues know that Jared does these things in physiology that involves laboratory science, and it's this strange stuff that we ecologists don't have much use for; but it doesn't do him any harm because Jared still does his bird work, so we have no prejudices against it. In short, my colleagues have been very tolerant. I think what's also helped is that, in the last dozen years, as my science has become more synthetic, and as a lot of the science that I've done has put together physiology and evolutionary biology, it's become clear that I can get a lot of mileage, intellectual mileage, out of working in disparate areas.

LEROY HOOD
was born in 1938 and was Bowles Professor of Biology at the California Institute of Technology. He is now at the Department of Molecular Biotechnology, University of Washington, Seattle.

CHAPTER 5

Scaling the heights

Leroy Hood
Biotechnologist

BIOLOGISTS can be rather snooty when it comes to technology. While they value new techniques, inventing them doesn't carry quite the same cachet as a clever experiment or idea.

It's a prejudice Leroy Hood does not share. An immunologist, he has played a major role in developing the instruments for the automated analysis of genetic material and proteins. It is the technology on which some of modern biology now depends, and underpins ambitious projects such as the international effort to try and map the entire human genome.

When we talked to Hood he was still at the California Institute of Technology, running a lab recognized as one of the largest and most innovative centres of biotechnology in the world. Apart from a recent incident when two of his postdocs were accused of fraud, he has gone from one success to the next.and has now set up a new and even more ambitious department at the University of Washington, Seattle. How does he do it?

We sat down to talk in one of his two immaculate offices in a modern building arranged around a courtyard of pools and fountains. How did his early training in medicine lead to his subsequent career in high technology?

It came about, basically, because I'd decided that human biology was going to be my real interest and as an undergraduate at Cal. Tech., a very technically oriented institute, I decided my background in things human was very bad, so I decided to go to medical school. The idea, initially, was just to do the first two years so that I could get the general survey courses, and then switch over and take a Ph.D. But at Johns Hopkins University where I went to medical school they had an accelerated programme, so I got through in three years and then went directly into the Ph.D.

'So you never really intended to be a doctor?'

I never intended to be a physician whatsoever, although I must say, at the end of my medical school tenure I had everyone come and say 'You can't waste all this time. We've spent all this effort educating you, and here you are, just going to go back and get a Ph.D.' But I said ' I was straight with you in the beginning, and this is where I'm going.'

'Why biology?'

I think probably the most determinative event in choosing biology was my senior year in High School. A chemistry teacher came to me and said 'Look, I have to teach biology. I don't know very much about biology. Why don't you help me by teaching it?' So what I ended up doing was reading articles in *Scientific American* magazine. Each lecture was a different *Scientific American* article. And by the time I'd given twenty-five or so of these lectures I was really convinced that biology was absolutely incredible. I'd kind of leaned that way before, and this teaching experience really crystallized for me the beauty of the questions that one could ask, and the fact that even at that time you could see about new ways of approaching these questions.

'Now, as I see your career, it's really got three strands. There is the medical side, but there's also a very strong science and technological side. Do you see them as different?'

No. My view of biology is that it must be a constant interplay of asking fundamental questions and then developing the tools that you must use to get the answers to those fundamental questions. My career in many ways, as I see it, has been posing a question, developing a tool, using it, and then going up to the next barrier, posing a question and then trying again to develop tools that would get us by that barrier. So I see what I do as this beautiful fusion between biotechnology on the one hand, and leading-edge biology on the other hand. The two are absolutely integral.

'But you are unusual in that respect. I don't think most scientists develop new technology. In fact there is an old split between pure science, and applied science.'

That's absolutely correct. But in a sense biologists have always developed techniques, although in a very *laissez faire*, cottage-industry kind of fashion. What convinced me that we have to be a little more organized about this, and bring in all these other disciplines to biology, is the complexity of modern biology, the complexity of the questions that we have to deal with. We can't just study a single gene or a single protein and how they function; we have to understand collections, complex systems. And to do that we've got to develop all sorts of new tools.

'But you went into immunology first.'

Correct.

'Why?'

I think the major reason was, number one, it was clear at the time that there were going to be beautiful systems for exploring the fundamental properties of the immune response because we could get a hold of the cells, and we could get hold of tumours of these cells, and we could get their products, in that particular case, the antibodies. But, equally important, it's a system that poses absolutely fascinating biological questions. I remember in my first year at medical school giving a report, back in 1960, on theories of antibody diversity and I almost then decided that immunology was what I wanted to go into.

'Very early on you started developing new techniques?'

That's absolutely correct. One of the fundamental opportunities for thinking about how antibody diversity arose was to explore the basis of this diversity, the antibody molecule. That meant developing the protein chemistry that allowed you to analyse the sub-unit components of protein molecules, the twenty different letters of the protein alphabet, and known as protein sequencing. It became obvious to us early on that we had to develop better and better machines for doing this if we wanted to do it more rapidly, and if we wanted to do it using smaller quantities of material and so forth. This was an imperative for getting us into our first technological effort, which was developing an automated machine for sequencing proteins.

'But are you then more interested in developing the techniques than actually solving some particular problem in molecular immunology?'

They're a chicken and an egg, you can't really separate the two. If I want to solve a particular problem, the only way I can do it is to develop a new machine. I see the discovery of the antibody diversity in the development of a protein sequenator that let us approach this question in a very powerful way. I see them kind of as yin and yang in moving biology ahead.

'I'm sorry to pursue this point about technique, but there are people who say that the trouble with the field—not necessarily your field, but certainly part of my field, developmental biology—is that people don't think any more. There are all these enormously powerful tools, some of which you've developed, and they simply apply what's there. So it's almost become very high-class cookery, just following recipes. As a result, a lot of the intellectual excitement, they argue, has gone out of the field. Do you think that's true?'

I would agree with that point of view completely. One of the potentially detrimental consequences of giving people very powerful tools is that they think the tools can do the thinking for them, that they can plug them in and do more and more of the same kind of thing. That isn't what science is all about. Now, sometimes you do really useful things as you go along cataloguing things, and I'm sure you've heard of the division of scientists into the librarians who do lots and lots of cataloguing, and a few people who are out there at the cutting edge, pushing things in new directions. My own view is that probably both are necessary. Theories are great, but you often need a lot of information before you can sort out the reality from illusion. So it's very complicated, but anybody who takes a machine and does 50 000 protein sequences for the fun of it, I would question just whether that's adding much to science.

'What you enjoy, it's really inventing isn't it?'

I do like doing that, but I think my biggest passion is having a really clear vision about what can be done, and being able then to collect and attract and enthuse people about technical problems that can be solved. I have to say that my own abilities in engineering, my own abilities in computation, my own abilities in chemistry are reasonably modest. I'm conversant with these things, but the people who've really done it have been people that I have persuaded, who are experts in these fields, to come and focus their attention on these things. It was clear we needed a protein sequencing machine, and we put together a team that did it. In fact the machine that we developed ten years later is still a standard in the field, although we're about to develop another one that will replace it.

'But then it's a surprise to me that you like science fiction.'

Oh yes, yes.

'Is that part of your image of the future?'

Science fiction has fascinated me because I like to see how people think about alternative realities and I like to see if I could have done the story better myself. In fact, when my children were growing up, one of the things I did was create this mythical science fiction character called Harry Golden who always got into these enormous difficulties, and then the children had to problem-solve about how he got out of them. They are now both superb analytically, and I'm absolutely convinced that the Harry Golden stories had a lot to do with that. So, it became kind of an early way of life with the family and so forth.

'The stories you told your children, clearly, you think, had an important impact on their mode of thought.'

That's correct.

'What, then, in your own childhood influenced your mode of thought?'

I think there were two things. The first was my father was an engineer, an electrical engineer, and I think he always wanted me to be an engineer rather than a scientist.

'In a way he's turned out to be right.'

He's turned out to be right. He, for example, used to take me to the courses he taught his men. He would involve me in these courses and delight in showing those men that his son could keep up with all of them. But I think in some ways probably even more important were the opportunities that I had in the form of three outstanding High School teachers in a very small town in northern Montana. There were 150 kids in the whole High School, and what was unusual about these teachers is they all cared about students as individuals. They spoke to you as a real person, and they, in different ways, gave me an alternative vision of things I might think about doing that went far beyond anything I could have imagined. Even more than that, they were enormously positively reinforcing.

Maybe this is a consequence just of small towns, but if you're good in a small town, everybody knows it and you have an enormous self-confidence about, gee, this is wonderful, I can really do all these things. Then when you go out into the real world, if you're good you never lose that confidence, right, it carries you through all the other kinds of things that happen. What is interesting is that I once had a conversation with one of these fellows who goes around and specializes in making places more efficient, and I said 'Well, where do you get people who can analyse all these different things?' and he said 'I get them from only two places. I get them from small towns . . .' And it's the same thing I said, 'I get them from small towns where they were the best all the way through, and they never lost that confidence. Or, I get them from places like New York City where, if they're number one, they know they're the best. I never got anybody that was any good from suburbia. That's middle America. There isn't a push to excel; there isn't the positive reinforcement.' And I subscribe to that point of view.

'You operate on a big scale.'

That's correct.

'Why?'

Well, perhaps I was a little glib about saying that's correct. The science I do I wouldn't put in the category of big science of the super-collider type or of the space station type. I would say that rather than on a big scale we operate on the scale that's necessary to put together all of the interdisciplinary components that our science requires.

'I should just ask you, how many people are there in your group? How many people are you really responsible for?'

Well, the number of scientists we have in the group is roughly thirty-five right now, so it's certainly a very large-sized group for molecular biology, but it's broken down into many smaller groups. So, we have a group that does synthetic organic chemistry, we have a group that does mass spectrometry, we have a group that does various aspects of computation and another group that does chemical microsequencing. And all of these groups—and their numbers are never more than three or four typically—all of these groups then have to be blended together into an interacting set where, given a particular problem, we can ask the question, 'what is the best technological approach to try and solve it?' But at the heart of doing science, as you realize, is an independence, a realization that everyone has to have their own thing and, of course, that is the challenge. In trying to put together an interdisciplinary group, it's the challenge of balancing the imperatives of these really difficult problems with the requisites of independence, of originality and creativity, that are functions of individuals.

'Is there not a danger, though, with working on a large scale, that fraud becomes more likely. This has received an enormous amount of publicity generally, and also in relation to your own lab. Do you think that some of the problems of fraud come from working in large groups, or does it come from the intense competition?'

Oh, I think it comes from the intense competition. If you look at the fraud cases in the States—and there are a lot more than have ever been reported, quite frankly—the ones that get a lot of publicity are those that have come out of David Baltimore's lab or my lab, and in thinking about it—and I've obviously thought a lot about it over the last few years—I think it really is exacerbated very much by the intense competition for jobs. I mean, once you finish a postdoctoral fellowship, you've got to get a good job, you've got to be able to get a good grant, before you can think about being awarded tenure. There are enormous pressures on individuals now that just didn't exist when I was coming up. When I got my degree I went out and did a postdoctoral fellowship; then wrote my first grant up. I wrote it up in two or three days and I had no concept that I wouldn't get it, I mean it was just an automatic kind of thing. And that's changed. Grant writing, now, is a one or two or three month ordeal for most people, and they may have to go back three or four times. So, somehow we have to think very seriously about how to humanize what science is unfortunately becoming. But superimposed on top of that there is another thing that is very worth doing, and that is giving people very explicit statements in their early training about what science is about, and what's expected of them, and what ethics is, and what's appropriate and what is inappropriate. In a number of fraud cases it's easy for people to say, 'Well, I didn't really think that was fraud' or 'This was just a short cut.' If you've made it very clear from the beginning, using examples—I think examples are a marvellous way to illustrate what fraud is—then people will be sensitized in a way that will make them think twice about the short cut kinds of things that can turn out to be fraud.

'Were you shocked when you discovered it in your own lab?'

Shocked, depressed, horrified, overwhelmed. I mean in eighteen years, certainly, never had there been any kind of a questioning of integrity, any hint of anything like this happening, and then two individuals right together. It was certainly one of the most difficult and depressing experiences of my life to sit through and sort through this thing. And that was also the first time I ever came up against the kind of limits that were determined by things over which I had no control.

'With hindsight, do you think you could have avoided it?'

I think with hindsight we could have avoided it. I mean we're doing two things now. When new people come into the lab I sit down and talk about this. I go through the implications and how hard it was, and the rationalizations that these people gave, just so people understand those rationalizations aren't acceptable, they are fraud.

'Can you tell me what those rationalizations were?'

The rationalization that one of the people gave was: look, I knew what the answer was, I'd had difficulty getting the answer, so I put it together because I was sure and confident that this was in fact the answer. He claims he didn't really intend to deceive or any of this kind of thing.

'And the other one?'

The other one, the other fraud, really centred about a trivial thing, manufacturing some control so he wouldn't have to run something again. He'd done them before, he knew exactly what the answer was, but he took a short cut because he wanted everything to be nice and on the same gel and this was an easy way to do it. I mean it was just an abysmal kind of thing. But in both of these cases I have to say there were also a lot of data missing. In one case the person claimed it had been stolen from him, in the other case the person had just moved to a new place and he said, 'gee, I didn't think about it, I threw it all out.' So, we don't know for a lot of data whether there were other things or not.

'It's interesting that you say it's really competition for jobs rather than competition with other labs.'

I think these two scientists viewed it in a very personal way. They were talented, they were intellectually gifted, they were extremely hard workers in both cases, and I think each of them really wanted to have, carve out, an exceptional career, and the way they saw for doing that was to take lots of short cuts so they could produce more work.

'What about Cal. Tech., you're about to leave Cal. Tech.?'

I've been here twenty-one years, it's been a spectacular place for me to grow scientifically, and there is, I think it's true to say, no other place that I could have evolved the particular type of career that's fused technology and biology so beautifully. And that is not only because of the talent of the students and the quality of the faculty, but because Cal. Tech. is a small place. You rub elbows with a physicist, a mathematician and a computer scientist, it's easy to get to know these people, and making the decision to leave was probably the hardest thing I've done in my life.

'What decided you to leave?'

The major thing was I wanted to do things that I couldn't do here at Cal. Tech., and they've revolved around the science. Putting together this large interdisciplinary group that I have, in order to stabilize it—and some of these people have been with me for five or six or seven years—I needed to be able to offer them real faculty positions, something beyond just advanced postdocs. Cal. Tech.'s point of view was, look, why don't you let these people turn over and every three or four years you can bring in new people. But it isn't that simple because the talents we're talking about don't come from a three- to four-year postdoc. So, the choice I had was then to very comfortably remain at Cal. Tech. and continue to do terrific science in this wonderful environment, but to accept the limitations of the continual turnover, or to move to Seattle and create a department that can then legitimize the appointments the good senior people can get, and stabilize this whole entity and create a division, a department, that will have as its central view this view we've talked about of combining science and biotechnology. I think the other thing that persuaded me Seattle was a particularly unique opportunity was, on the other hand, some really good young departments. For example, the computer science department out there is absolutely wonderful, they're young people who are really enthusiastic about getting involved with us. Now, here at Cal. Tech. we have a small computer science department, it is excellent, but they have all the money they could ever want, they have terrific problems, so to get their attention is more difficult. Going to a place that has younger people, where the department is trying to build itself up and become very much better, has a sense of wanting to join the new ventures.

'One of the reasons that you're going to Seattle, though, is that you want to climb mountains more?'

That's another aspect of it too, yes.

'Do you see your science and mountain climbing as rather similar challenges?'

Oh absolutely, and in fact mountain climbing in some ways is even better than science because it combines both enormous intellectual challenges—it is a real technical problem to solve every new mountain that you'd like to scale—and just this beautiful, physical exertion, pushing yourself out to the limit. In fact it's a wonderful complement to science because in doing that you can forget about science for a week, or ten days, or two weeks, however long you're out mountaineering, because you focus on the opportunities and the challenges and the excitement there. So, I find my mountain climbing absolutely essential to the renewal that lets me go back to the lab and be all excited and start all over again.

'One last point, you quite often use the word "beautiful" about science, what do you mean?'

Well, I think it's a part of my natural enthusiasm for everything, but what I've been impressed with in science over my twenty-one years is the absolute conflict between, on the one hand, as we come to learn more and more about particular biological systems there is a simple elegant beauty to the underlying principles, yet when you look into the details it's complex, it's bewildering, it's kind of overwhelming, and I think of beauty in the sense of being able to extract the fundamental elegant principles from the bewildering array of confusing details, and I've felt I was good at doing that, I enjoy doing that.

SIR DAVID WEATHERALL
*was born in 1933 and is Regius Professor of Medicine,
University of Oxford, and at the Institute of
Molecular Medicine, Oxford.*

CHAPTER 6

Clinical science

$\backsim\!\!\!\backsim\!\!\!\sim$

David Weatherall
Clinician and molecular biologist

SIR David Weatherall belongs to a very small group of outstanding doctors who are also outstanding scientists. Neither the conventional image of the caring GP, nor the suave Harley Street specialist, is easy to reconcile with the stereotype of the cold, analytical laboratory scientist. This implies the question of whether, in the United Kingdom at least, there is a genuine difficulty in practising good medicine and doing good science at the same time. As Professor of Medicine at Oxford University and Director of a highly successful Medical Research Council research unit, David Weatherall does do both: how has he managed it?

As a haematologist he is concerned with diseases of the blood, and has been at the forefront in applying the techniques of molecular biology to understanding the basis of inherited blood diseases. One such disease is thalassaemia which results in insufficient haemoglobin being synthesized in the red blood cells, so impairing their ability to carry oxygen. Haemoglobin is a protein and, like all proteins, its structure and production is coded for by genes. A defect in the protein can be traced to a defect in the DNA of the gene. The techniques of genetic engineering are increasingly allowing such defective genes to be detected prenatally, and opening up future possibilities of gene therapy—the replacement of the defective genes by normal ones.

Given the anxiety created in many people's minds by the prospect of tampering with our genetic make-up, and the ethical issues that such possibilities inevitably raise, Weatherall is working right at the point where, in theory at least, he could find a contradiction between his position as a doctor and as a scientist. Does this in fact happen?

The nature of the dual role raises more pragmatic questions, too. The demands on Weatherall's time are awesome. As well as seeing patients and running a lab, he has to take on board administration, the demands of the National Health Service, teaching, fund raising and innumerable committees. He also edits and oversees a number of large and successful textbooks. How does he fit it all in? His manner and appearance give no hint of the pressures upon him, or of his achievements. With his tweed suit and pipe, he could be mistaken for a moderately successful farmer.

We met in his lab at Oxford and decided to start at the very beginning with the question of why he went into medicine in the first place.

I've no idea. It's an awkward question that you're always asked when you're interviewed to go into medical school. I just always wanted to do it. I think I had a very early desire to be a plumber, but after that, it was always medicine, right through childhood, for no apparent reason.

'When you went into medicine you didn't have any idea of doing research, or did you?'

No, none at all. In fact, I didn't have much idea of doing research right through medical school. I trained at Liverpool, and there were one or two teachers, such as the physiologist, Rod Gregory, who were obviously doing research. He used to come in and lecture straight from the laboratory, and one could sense the excitement. You felt he was not just teaching physiology, but making it happen. But I didn't have any thoughts that I might do that kind of thing myself.

'So when did you actually think of doing research?'

I never thought of doing it in a conscious kind of way. It was just pure chance. I saw a patient when I was doing my National Service in the army that interested me. There was a young child, in the military hospital in Singapore, who had a severe anaemia. This was a Ghurka child, and nobody had been able to sort out why she was anaemic. And for some odd reason I went along to the library and started reading a bit. It was just about the time that people were first noticing that these genetic anaemias, inherited anaemias, occurred in South-East Asia and, to cut a long story short, I used to potter down to the biochemistry department in Singapore University in my spare time, and we worked out that this child had one of these anaemias.

'What were you doing in Asia?'

Well I was in Asia, again by a series of chances. I had to do my National Service after doing a higher degree in medicine, but I'm terrified of aeroplanes, so I volunteered to stay and do it in the United Kingdom. So of course, the immediate reaction of the army was to send me as far as possible, and that was Singapore, and when I got to Singapore they put me in charge of the children's ward, and I bumped into this interesting child.

'Was it that interest in anaemia that made you a haematologist?'

Yes I think so. Once we'd sorted out that child, I went up to North Malaya to do the second year of my National Service, and found the whole idea of searching for these children in the population fascinating. They were obviously about all over Malaya, they'd not been previously recognized, and I got very, very fired up by all this.

'How then did you get into molecular biology?'

Well, another chance really. The Liverpool Department of Medicine had developed close links with Johns Hopkins Hospital in Baltimore, and it just so happened that when I was coming out of the army, there was nobody from Liverpool to take the Baltimore slot, so they wrote to me and said, 'Would you like it?' and I said, 'Fine'. And I wrote to Johns Hopkins and said 'I'll come and work on thalassaemia'. They didn't say anything, and when I got to Baltimore I discovered most of the thalassaemia was in New York, but we managed to get round that problem, and that's how I got into it. Or rather, that's how I got into decent genetics, and then after that I rather drifted into the more molecular end.

'You say, drifted in, but wasn't that where the action was?'

Well we were just starting to understand a little bit about the way in which human haemoglobin is inherited in the 1960s and, about the same time, the genes which control haemoglobin had also started to be understood a little bit more. We knew there were two, for example. The techniques were just becoming available which made it possible to study the production of a protein in the test tube, so that was how we started to get into the molecular end of this problem.

'And presumably blood was a good system because you have so much material, so accessible?'

Oh yes, there's no great surprise in the fact that the blood diseases were the first to be studied by the tools of molecular biology. It's always been a marvellous tool for protein chemists. I mean haemoglobin is served up almost a hundred per cent pure in red cells, and protein chemists, like Perutz, had used it as a model for studying the structure of proteins. Other protein chemists had used haemoglobin as a model for understanding how proteins are made in different bits, or sub-units, so we knew quite a bit about its structure, even by 1960. And once people recognized that human beings had variants of haemoglobin, we could study families, and work out the genetics of how those different sub-units of haemoglobin were controlled. So we were half-way there before molecular biology arrived.

'But here you were doing molecular biology, did you continue to do clinical medicine?'

Well in the four years I was in the United States I continued to do quite a lot of clinical haematology, and when I returned to this country in the mid-sixties I was probably doing about half and half; half research, half clinical medicine.

'How did the basic scientists perceive a clinician like yourself actually doing hands-on molecular biology?'

Well I'm not sure I've ever done really sophisticated molecular biology, but I think the answer to your question is that in this country they regarded me with extreme scepticism. In the United States, it wasn't quite so bad because there's always been a tradition of clinically trained people going into basic science. There was a little bit of eyebrow raising when an Englishman wandered into the biophysics department, and wanted to work there at Johns Hopkins, but it was not quite the same as if it had happened in this country.

'Why is there this peculiar tradition in this country?'

Very hard to say, and I'm afraid it still exists. Maybe part of it is because of the way that medical education is organized here. Somehow, in most clinical schools anyway, there's a major break between the pre-clinical science part of the course, and clinical science. It is perceived by students that there's not much of direct relevance taught in pre-clinical, basic science, and a great sigh of relief when its all over, and all the basic science books are thrown away when people go on the wards.

That's one reason. There are many other reasons, including finance and career prospects and so on, but I think the medical establishment in this country has never got to grips with the importance of well-trained clinical scientists having careers which give them the time to do good science with long-term continuity. We've said to the young clinicians, fine, you can have two or three years to do some science, but you must come back and do 'proper' medicine, and there's always been a feeling that you can either be one thing or the other, but nobody must try and be both.

'But is there a conflict, or a difference, between doing research and practising medicine?'

Well, that's a very difficult question. There are two conflicts. One is the simple pragmatic conflict of time. You need time to learn new techniques in research. And the other problem, of course, is that once you've trained yourself, or tried to train yourself, how far can you apply it. And there is another major difficulty because you can take clinical medicine at the bedside only so far. One can explain the scientific basis for some diseases, and the scientific basis of some methods of treatment, but then it stops, and the *art* of medicine comes in. Take a simple disease like pneumonia, or a genetic disease that you think you know everything about. The effect on an individual patient varies enormously. You've got their individual reaction to the disease, and the fact that a single-entity disease like pneumonia may react in many different ways to treatment. So in a sense you can take your science so far, then you have to cut off at the bedside because the patient doesn't want a long dialogue of what we don't know, he wants a positive approach.

'You're really saying medicine's more of an art than a science?'

Yes, I think so, and I suspect it always will be, even if we get down to the stage where we can take your last molecule apart, and explain most diseases at that kind of level. And that's because when you put that disease into the enormous complex thing that a human being is, it's going to react in so many different ways, and whether each of those facets will be explainable on pure scientific grounds I very much doubt. Take just a simple example, an inherited anaemia that makes the average person, perhaps, feel a little bit tired most of the time.

You know very precisely what's the matter with the gene, and that these people are tired because they've got too little blood. But the individual's reaction may vary enormously. For example, we know nothing about the complex interactions and individual genetic variability of the numerous adaptive processes that come into play to compensate for anaemia. And peoples' psychological make-up varies so enormously and for many different reasons. It is easy to ascribe symptoms to mild anaemia only, on further delving, to find that they are actually suffering from minor depression. Patients' home lives, the support that they have, and the reaction of their relatives and employers to their illness have an enormous effect on their individual response to disease.

'So what then makes a good doctor?'

Gosh, that's even more difficult to define. But in the context of what we're talking about, I don't think that because somebody is trying to be a scientist as well, that it necessarily makes them into a bad doctor. That's been another myth which has been fairly popular in this country.

'But it doesn't necessarily make them a good doctor, that's what you're saying?'

No it certainly doesn't. I think to be a really good doctor there must be a major self-critical element which is probably helped very much by good scientific training, and probably at least a period in medical research. But after that there's a number of rather undefinable elements. Probably the most important one is the ability to communicate well with people because unless you can do that, you're never going to be able to sort out the multi-facets of any disease.

'You, although you're a very distinguished physician, don't give the feel of a Harley Street doctor. Is that conscious?'

You mean do I make an effort *not* to appear like a Harley Street doctor? I don't think it gives me the slightest trouble not appearing like a Harley Street doctor. I'm not sure what the Harley Street doctor image is . . . I presume you mean somebody who's excessively well dressed, extremely smooth speaking, and perhaps has one eye on the bank up the road.

'And also supreme confidence.'

Supreme confidence—yes—confidence is a difficult thing to evaluate in medicine. There's no question, I think, that a good clinician needs confidence, or he must seem to be confident at the bedside. Now you might argue that somebody who is bathed in basic science is just the opposite, has enormous caution, and that does cause some kind of problem. But that's what I mean by there being this kind of hinterland between science and clinical practice, of having to be able to switch into the art from the science. The hardest thing to teach medical students is when, in managing a particular patient, that moment arrives when genuine scientific explanation and method no longer apply and kindly empiricism must take over. Maybe having a long career in clinical science makes one a tiny bit more humble, though. I don't know. I think the old Harley Street image is largely going out of British medicine.

'But consultants still do have a curious sort of power within the clinical situation. There is no one who ever really questions their judgement. Now that's quite different from being a scientist.'

Yes, you're absolutely right, that is very different from science. And that certainly used to be the case in medicine, although it is beginning to change now. When I was medical student I was almost put off continuing because one questioned nothing, and the consultants' views were the final word. In fact, it was rumoured that my professor of medicine had never made a diagnostic error. And people seemed to genuinely believe this. Incredibly unhealthy for young people. Now things have improved, I think. One of the problems is that until, what, ten, twenty years ago, all medical practice was totally empirical, and therefore this kind of aura of mystery surrounded the profession. As science has crept in, this is gradually clearing, and people are questioning a lot more. The attitude you refer to, what we're seeing now, is probably just a remnant of that kind of era. You don't see it nearly as much in the United States, and I notice it's improved a lot here in the last five or ten years.

'But isn't it true that quite a lot of a clinicians are really quite frightened of science?'

Well there's no question about that, and that may be one of the reasons why this kind of barrier against science is put up, and it's considered to be unimportant. It's probably not surprising, particularly in recent years as science has got so complex. But I think it still goes back to the problems of the training of clinicians in this country.

'Do you care that your research leads to practical results?'

Yes . . . I do. It's an interesting exercise running a research group with a mixture of basic scientists and clinical scientists, because there is a tendency, even in a group like mine, to be doing interesting science for its own sake. What we have to do, I think, is not be too hung up on immediate practical results. But it has given me enormous pleasure working in the genetics of blood diseases to see how, by working out the molecular defects over the last few years—which at the time didn't have any immediate prospects of being useful—within about five years they were being used in the clinic for prenatal detection of these diseases, and I think anybody who works in science in a clinical setting must always have a long-term goal of something useful.

'But do you get pleasure from basic science too?'

Oh enormous pleasure, yes. That's the trouble, everything's so interesting. I mean in our particular field, although those practical things are always in one's mind, there's a tremendous kind of temptation to shoot off into problems of evolution and population genetics, because we can begin to answer some of these things now. You could become a kind of gorgeous amateur at everything.

'You yourself do an enormous number of different things, don't you?'

Well I try to, yes.

'Where do you get the time?'

I don't know. I mean I probably work ridiculously long hours by some people's standards.

'How long?'

How long? Well I normally get in at about 7.00 in the morning, when you get that marvellous hour's peace before people start arriving, and will spend the morning, hopefully, on the clinical administrative side. Maybe the average afternoon, get a little time with the research folk. Wander home at about 7.30, and then I suppose I have to admit that perhaps five evenings and a bit of time at the weekend I do the writing and stuff, between perhaps ten and twelve or one in the morning, which is my best time anyway. So it's a rather stupid existence.

'But you must be very well organized. Have you learnt to do this, or does this come naturally?'

Oh good gracious no, no. I've learnt a little about organizing my life, but I lived in total chaos probably until about ten years ago, and then just relative chaos for the last ten years.

'What was the switch?'

Pure necessity I think. It was either going to be an early grave, or get a bit organized.

'Doctors, in the clinical situation, have enormous power. Where do you think, or how do you think, that certain clinical decisions should be taken? Do you think the patients and the public should be involved in this, or should this be left entirely to the medical profession?'

I think until recently, the medical profession has probably had much too a paternalistic view of its interactions with patients. On the other hand, when you've got a sick patient, or say, the parents of a child with a serious genetic disorder, I think a certain amount of paternalism probably is necessary. A lot of these people will come to you and you can sense they're asking you to take part in the decision. You can see the look of relief when they ask the question 'what would you do?' So I think that a certain amount of paternalism is necessary. But I think in the past we've taken it far too far, and there's been this God-like attitude taken by the profession. Now, particularly in some of the things that are going to happen in the none too distant future in terms of genetic manipulation and so on, we've got to open a debate with the public. It's got to be a kind of two-way affair.

'How do you see that developing? Do you see that as a real set of new ethical problems, genetic engineering?'

I think most of the ethical problems have been around for a long time. What genetic engineering's done is just made it a lot easier to do things like identifying disease in the fetus and so on, so they're becoming much more common ethical problems. But when we start manipulating human DNA, putting genes back and so on, we will be entering new areas, and there has to be some kind of debate with the public. How we get that going is not clear at all. I think the scientific community has to make every effort to write in a popular way—in an understandable way—and through the media, radio and television, we've just got to get these messages across, and let people debate them.

'Do you think many of the anxieties people have about genetic engineering are realistic, or are they just fantasies?'

I think that the majority of them are probably unreal. If one looks forward to the immediate future anyway, we're not going to do anything which is not being done already really. I mean people are very concerned that putting new genes into people is a departure. But if we're putting those genes into body cells—not into the germ cells, where they'd be passed on to the next generation—we're fundamentally doing no different than we do with organ transplants. I think the worry is that as we get more and more sophisticated with the neurosciences and start to learn more about behavioural genetics, the temptation might be to try to change human beings. We're light years away from that at the moment, and my guess is that the number of the genes that will be involved in complex traits will be so many that it would be highly unlikely one could do anything like that anyway. It's interesting to speculate, but I think we've got much more pressing problems in medicine and biology than worrying about that.

'But do you think there are areas that shouldn't be researched, as it were, areas of forbidden knowledge?'

No. Because I couldn't draw the line at where one might stop.

'Do you think there's any problem then, for example, with sequencing the human gene? A lot of people see that this is going to raise a whole new set of moral issues that, if you knew someone's total genetic constitution, then you would make all sorts of predictions about them.'

I think that there are concerns, although I think the major ones that have been raised so far have been exaggerated . . . if one is applying for a credit card, will you require a total genetic profile in the future, that kind of thing. You've never felt terribly badly done to if an insurance company, or even your employer, wants to know your blood pressure, and the idea that we'll have a total genetic profile for everything, including whether we're bad tempered on Friday evenings, seems to me such a futuristic myth as to really be not worth talking about. What we will have, I suspect, is some very clear ideas about our likelihood of getting what are the common diseases of Western society, if you like. We'll certainly know, be able, probably, to define sub-groups of people who are more likely to have heart attacks at fifty rather than seventy, and probably the same for major psychiatric disease. Now, is it wrong to know that? My feeling is that we need to know that very badly. If one wants to prevent heart disease, at the moment we can vaguely try to make the population run around a lot, and eat less fat, but what we really need to know is what particular sub-groups are at particular high risk, so that we concentrate our preventive efforts on those. So the good that will come out of the gene-mapping projects seems to me to far outweigh the rather vague risks.

'Why do you think people are so anxious about genetic engineering?'

Oh, it's about as close to the soul as you can get, isn't it, with our genes. I think there's an enormous kind of mystique around it. And then I think people are genuinely worried about what they perceive as meddling with future generations. If you'd asked me for one area that I would not be . . . not knowledge that I didn't want to have, but a particular area of scientific activity that I would not encourage, it would be the attempt to insert human genes, or foreign genes, into human fertilized eggs with a view to them being expressed in future generations.

'Why not?'

Well that is changing the genome. It is giving our next few generations very little choice about what happens to them, and it seems to me that if we're going to do that, we would need a much broader debate on how far we want to go with altering the human species. It seems to me that that is a completely new step in practice, and I'm probably too middle aged to want to contemplate that at the moment.

'Presumably, being a doctor is stressful. Do you find it more stressful than doing research, and does research, as it were, give one a relaxation from the stress of being a doctor?'

Well it's very interesting. I think that the two activities are so different. In science I don't think there's anything more stressful than being somebody who's just had a paper rejected by the journal *Nature*, or has just got to the very end of a super series of experiments to find the whole thing was an artefact. That's quite as stressful in many ways as a doctor facing a family who've just lost somebody. It's just a different kind of stress. I find it really quite exhilarating, after a particularly bad ward round, to go back and see the research group because the stresses are just different.

'And maybe if one's going badly, hopefully the other one's going well?'

Yes, usually they're both going badly, but one does get the odd week when they balance each other out.

'What would you say your skills were particularly?'

Gosh, that's a terrible question. Probably just to have a reasonable kind of modicum of good northern common sense in most of the things I do.

'And your failures?'

Failures?

'Well, your weaknesses, I should say.'

Oh gosh, that would take too long. I think probably a major weakness is just total impatience at the kind of slowness of the kind of vast meandering systems I have to work with. For some odd reason, maybe it would be the same in any profession, I think the university system, and the Health Service system seem to have an inbuilt inertia which just sends me home some nights almost to the bottle, but not quite.

'On those few occasions when you have some time, when you're not working, how do you actually relax?'

Oh well, the one thing that totally relaxes me is music, so there's an enormous amount of late night music played, mainly as a listener, but when people are in bed, and nobody's around to listen, I tinkle around on the piano, and get an enormous amount of relaxation out of that. That's the one thing that keeps me sane.

SIR JAMES LIGHTHILL
was born in 1924 and was Provost of University College,
London, where he is now an Honorary Research Fellow.

CHAPTER 7

Swimming with the tide

James Lighthill
Applied mathematician

APPLIED mathematics does not have the glamorous image of pure mathematics. In the United Kingdom, at least, the very idea of application seems to make a subject somehow less imaginative, less innovative, less intellectual. Yet it is the application of mathematics that gives it its real power. Since Newton, applied mathematics has transformed our understanding of the world, and allowed us to manipulate our environment. It is applied mathematics that has made it possible to put a man on the moon, and planes through the sound barrier.

Sir James Lighthill has been described as Britain's greatest applied mathematician this century. He is also a renowned linguist, and polymath, with a wide knowledge, not only within science, but in the humanities too. He is, in addition, a distinguished administrator. At the time we interviewed him, he was Provost of University College, London. Earlier in his career he was director of the Royal Aircraft Establishment at Farnborough. For most of the years in between he was Professor of Applied Mathematics at Cambridge. It is quite unusual to find so brilliant a scientist dividing his life between research and administration in this way. Why has he done so? Are the satisfactions he derives from each activity quite different, or do they have more in common than one might expect?

Lighthill has made major contributions in many areas of applied mathematics, and especially in fluid mechanics, the study of the behaviour of liquids and gases. His research has embraced areas as diverse as subsonic and supersonic flight; jet noise; blood flow; fish and invertebrate locomotion; wave-action and ocean currents. He is also an enthusiastic swimmer with a penchant for swims of epic proportions.

His ascent to the higher reaches of the mathematics world was rapid. At Cambridge he studied under the legendary pure mathematicians, Hardy and Littlewood. After a time at the National Physical Laboratory, and a period back at Cambridge, he went to Manchester where at the age of only twenty-six he was to become a professor. But how early had his extraordinary abilities shown themselves? Had he, I wondered, been what one might call a child prodigy?

Well I always did well at mathematics at school, and I got a scholarship to Winchester at twelve, and I worked with a number of brilliant young boys at

Winchester who became famous, like the physicist Freeman Dyson, and the two Longuet-Higgins brothers, and we all conspired to learn a bit faster than we were supposed to. Dyson and I both took scholarships to Cambridge at the age of fifteen, and we were wisely restrained, I think, from going up to Cambridge at that age. I always, looking back, am pleased that I concentrated a good deal on the humanities during the extra year, and had all the advantages of learning languages and history and law and so on, before going up to Cambridge as a mathematician. At the time I'd rather have concentrated on my mathematics, but looking back I see that it helped me in a lot of things I've done later in life.

'But were you a prodigy in mathematics?'

Well, I think that that's implied by what I've said, in a way. We went ahead very fast in mathematics, and then when we went up to Cambridge we found we didn't really need to go to any of the undergraduate lectures, and so we went entirely to post-graduate lectures. And this was very advantageous because it was during the war, between 1941 and '43, so there were very few students staying on for post-graduate lectures. Dyson and I essentially had the benefit of some of the great people in mathematics, like Hardy and Littlewood and Besicovitch, almost to ourselves. So we learnt a great deal from them. They were pure mathematicians. In applied mathematics, almost everyone from Cambridge was taking part in the war effort, so we got up to a good level in pure mathematics, and we took simultaneously the main degree examination, and the post-graduate examination at the end of our two years, and then went into the war effort.

'James, do you think that mathematicians are, as it were, born, or is there some family influence that, as it were, helped you with your talent?'

My father did encourage me. He was an engineer, and he encouraged me at the age of about three to make rapid progress with mathematics, and so at my nursery school I was already asking to be pushed up a few forms, and successfully winning the right to do some more advanced mathematics, and so it went on later on in life.

'But do you think it's that family influence, or, as it were, do you think there are a set of genes that really made it possible ?'

It's very difficult to be sure, but certainly mathematics always came easily to me. At the time I felt certain that I would become a pure mathematician, although I suppose the fact that I was interested in a number of other things might have been a warning that it would turn out differently. But when I went into the war effort I immediately started research under one of the people who would have taught me if he'd been at Cambridge, a great applied mathematician called Sydney Goldstein, and he was a superb research supervisor. When I went into the National Physical Laboratory, Professor Besicovitch, the famous mathematician, wrote to Goldstein and said, 'we've got this good pure mathematician, please don't ruin him.' But Goldstein used to tell this story in later life with the addition, 'but I ruined him' . . . [laughter]. In other words, turned me into an applied mathematician.

'What's the distinction between applied maths and pure maths?'

Well . . . applied mathematics is using mathematics to solve problems in the real world. It really all started with Newton, because it was Newton who demonstrated, with his *Principia Mathematica*, that a whole area of experimental knowledge could be expressed in terms of laws which were mathematical in character. And this was really the first realization that the behaviour of the physical world can be described in terms of mathematical equations. So Newton achieved this bridge between the physical world, and the abstract world of mathematics. It was not at all obvious that such a bridge existed, but its correctness has persisted in the subsequent history of science. Well, of course, once that happens, then it becomes possible to treat practical problems by thinking about the mathematical laws governing them. Many applied mathematicians work in areas where the fundamental mathematical laws are well established. Nevertheless, a great deal of mathematics has to be done to apply them to a practical problem, like how to help jet engines become more powerful, without becoming noisier.

'Is the distinction between applied mathematics and pure mathematics one of motivation, that you're really interested in the external world, rather than the mathematics?'

That's right. And one is concerned always with building the bridge soundly. You have to establish the means of communication between the experts in the practical problem, and the experts in the mathematical problem. You have to use the mathematical theories to throw light on the problem, but then to make any effective contribution to the practical problem you have then to translate what you've done into the language appropriate to the practical problem. And I found that I had to do this whether I was working in engineering, or later in oceanographic, or biological problems, and so on. There's a problem of translation between the practical laboratory world, and the world of mathematics, and so there's always that extra dimension which I think makes applied mathematics fundamentally more interesting.

'Is the mathematics that you use in this application the sort of mathematics that Newton used, or are there a whole lot of new mathematical techniques?'

Oh quite new. Newton used the mathematics available in his day, and applied mathematicians in each subsequent period have used the mathematics available to them. I had the advantage of being taught by Hardy and Littlewood some very advanced mathematical analysis. It was advanced in their day, it wouldn't be advanced now. And the attractive thing about it is that Hardy and Littlewood simply had no idea that what they were teaching me was of any use at all. Hardy is famous for having said that the mathematics he did was of no practical use, but I found that the foundation that he gave me was just what I wanted to tackle the difficult problems of supersonic aerodynamics and jet noise and so on.

'What was the problem that really made your reputation so young?'

Well, it was these two things. It was a lot of work on supersonic aerodynamics, particularly recognizing that supersonic aircraft shapes could have quite low drag if they were really slender, and you can see a direct line between that and applied mathematics and the sort of shapes that ended up in Concorde, for example. And then the fact that I had the opportunity of moving into this exciting jet noise topic, I think, added to my reputation at that time. There were a lot of extremely interesting questions involved with a subject like aerodynamics which had been going for a very long time, and how it would change when shapes were going faster than sound, and you produced shock waves, and the shock waves interacted with the boundary layers attached to the surface of the wings and so on. It involved a whole lot of quite surprisingly complex mathematics. But, as I said, none of that is any good unless you can then communicate it to the engineers you're working with. And, conversely, you've got to spend a lot of your time in labs with engineers looking at their results, and then going away and working on how they could be interpreted. At Manchester we set up a very good mechanics of fluids laboratory, and I spent a lot of time there puzzling out the phenomena that were being discovered in the wind tunnels and the shock tubes, and then going away and trying to work out the mathematical analysis of them. So that was a very nice period.

'Now most of your work has been in fluids. Is there something about fluids that appeals to you?'

Aha, yes, I think so! I have a sort of general pleasurable feel about fluids and, of course, I'm very interested in flight, and although I worked entirely on aeronautical flight in those days, I subsequently did very comprehensive studies of animal flight—birds, bats and insects—during my later period in Cambridge, working with the zoology department there. And my hobby is swimming; I have a great deal of interest in the ocean—ocean waves, ocean currents, ocean tides—and so I enjoy observing all that when I swim. And then I have a fellow feeling for the swimming animals, and I've written papers about almost all varieties of swimming fishes and invertebrates, and quite a lot of work on micro-organism locomotion.

'Part of your passion for fluids is swimming.'

Yes, indeed.

'Do you swim a lot?'

Yes, I do a three-mile swim every weekend just to keep fit.

'And in the holidays?'

In the holidays I always do each year an adventure swim, which I do, partly because it's good for all of us to have an adventure every so often, but partly because when I was at Farnborough I was working with test pilots, and I was conscious that they were actually depending on the scientific work that was done; they staked their lives on the correctness of the science. I've done a lot of work on ocean waves and tides and currents, and I feel I understand them well enough to be quite prepared to swim in them, because with my theoretical knowledge, supplemented by an immense amount of experience in swimming in these conditions, I can swim safely, and have an exciting adventure in the process. So I do this, usually choosing swims where there are quite difficult currents to deal with. Sometimes swims round islands, sometimes swims between one island and another.

'Like what?'

Well one of my famous swims is the one around Sark which I've done five times, and one of them was during a south-westerly gale which was the one that actually caused the Fastnet disaster. So one needed quite a lot of nerve and stamina to complete that swim on that day, but it really was rather an exciting experience. But I've swum between two of the Azores which have quite a strong current between them. I've swum round an actively erupting volcano, namely Stromboli, and watched eleven separate eruptions from the side where you can see the volcano, where incidentally, the water is the temperature of a hot bath because that's the side the lava comes into the sea. And I've swum round Lundy, and my most recent swim was round Ramsay Island where there are exceptionally strong currents off the south-west coast of Pembrokeshire.

'Do you actually use your knowledge of waves and tides in order to do it?'

Oh enormously, yes. I mean during this Fastnet swim I was constantly having to sort of add up vectorially my swimming velocity and the current velocity, and the wave drift due to these very powerful waves. It was rather interesting. I was really having to swim at right angles to the direction I wanted to go in, which you often have to do, of course.

'I don't think many of us [laughter] would recognize that.'

And, of course, you meet seals and all sorts of interesting animals who have a fellow feeling with swimmers when you do these swims.

'It's very nice applied maths I must say. [Laughter]. Why did you, in a way, leave or give up some of your time for doing mathematics, and move into administration?'

I was challenged by being invited at the age of thirty-five to take on the directorship of the Royal Aircraft Establishment. I'd become well known for my work on jet noise and supersonic aerodynamics, and they thought that perhaps the RAE at Farnborough needed a more scientifically based direction, and I rather enjoyed giving it that during the next five years. And then I thought, well I don't want to go on doing administration for the whole of my life, so at that stage I accepted a Royal Society Research professorship, and subsequently I accepted the Chair at Cambridge, the Chair that Newton had originally held three centuries earlier. So I really enjoyed being able to do an immense amount of new research. I moved away from aerodynamics at that stage, into oceanography, and into biological applications of mechanics of fluids. And I thoroughly enjoyed Cambridge and so on, but towards the end of the 1970s I felt, well, perhaps it would be rather fun to run something again, so I was looking out. And then I was invited to be the Provost of University College, London, and that seemed a very exciting challenge, and it is where I've been able to use my great widths of interest. As an applied mathematician one hangs a sign outside one's door saying, 'Mathematics, Applied', and lots of people beat a path to the door, and so during a longish career as an applied mathematician you learn about an awful lot of different subjects, and that qualifies you in a way to be in charge of a very big, exciting, complex organization which covers a lot of subjects. And I have a lot of linguistic interests, and humanities interests, so it was a pleasure to be looking after a very fine faculty of arts, and since my biofluid dynamic interests have taken me into quite a lot of different medical problems, I've enjoyed looking after a medical school also.

'But do you actually enjoy the administration?'

At this level, yes. It's really rather an exciting challenge to look after a good place, and to keep a really good place good, and perhaps even to help it to get a bit better in difficult times. So I've really thrown myself into that.

'But have you discovered principles of administration? Do you see the problem a bit like applied mathematics?'

[Laughter]. There are some principles, absolutely. You can't run a big organization unless you're very interested in people. The essence of running a good organization is getting the best work out of people; helping their morale to be improved; picking the right people to do all the leading jobs. You've got to find people who are strong academically, and also have all the human qualities. It's vital in a big academic institution to devolve responsibility, because you have a whole lot of different specialties, and they will only be pursued well and effectively if you can find somebody for each major specialism who is going to give the leadership throughout that specialism, and will take the hard decisions about money and so on, taking into account his expert knowledge of the field. Not doing it as a business manager would, but saying, 'For my subject, say anatomy and developmental biology, what will do the best possible thing for teaching and research?'

'Do you see that as the core of your administrative principle?'

This devolution is very important. Incentives are important too. These leaders, heads of department, if you like, need to be given incentives to do the things which are best for the college as a whole, and for their departments. So I invent complicated formulas which will determine the resources they get, and the formulas are designed as incentives to cause them to devote a lot of effort to various things like bringing in extra money in support of their research and so on. One can use an applied mathematical approach to cost accounting, and to the determination of resources in a way that will make people adopt the best possible policies.

'Do you think running the university has been more like running a business, or is it still a purely academic exercise?'

The whole emphasis is on research and teaching, getting the best possible courses, the best possible research work, the best possible staff for both. My most important contribution, as I mentioned in a speech the other day, has been in the hundred or so professorial appointments that I've presided over during my period and, of course, a large number of other important appointments to readerships and senior lectureships. There's no doubt that the college today looks like it does, very strong academically in all the fields, more because of the professorial appointments that we've made during the 1980s than to any other single cause.

'So you don't see it as running a big business, the way perhaps the Government would like to see it?'

It's impossible to look at an academic institution's objectives otherwise than from the point of view of seeking excellence in teaching and scholarship. It's the right way, and you have to be concentrating on it all the time. Every day I'm involved in discussing with members of the college staff their academic objectives, and working out ways of promoting them.

'And you have, as you say, enjoyed being in a broader intellectual environment?'

Yes.

'Because you have an enormously wide range of interests.'

And I've been able to indulge them as Provost of UCL.

'Do you have total recall for everything you read and hear?'

No, no, no. It's not like that all. It's ideas I remember.

'And when you came to learn languages—you speak Russian and you read Russian—did you have difficulty learning languages?'

Well, I said that an applied mathematician is concerned with communication, and even translation between an experimental scientific situation and the abstract world of mathematics, so it's not surprising that I have linguistic interests. I think with every language you learn, it's easier to learn the next one. There are three languages which I read extremely fluently in the sense that I read novels in these languages in bed, and they are French, Russian and Portuguese, which I've thoroughly enjoyed learning. I think if you want to enjoy literature you are rather limited by confining it to the literature of one country. If you're going to live a long time you really need to enjoy the literature of other countries. My own strong belief is you can only enjoy them in the original. So it's worth learning a few other languages, and you know, one tries to learn a variety of different languages, and then you find that there's a small number, in my case these three, where you feel completely in tune with the people and the language, and the literature and you find it's a rewarding literature. So I feel I know almost as much about those three literatures as I do about English literature, on the basis of having read the material in the original.

'Is there anything that you actually find difficult?'

[Laughter] Well I only attempt such a modest range of things. [Laughter].

'That's not really true. But is there anything that you really find very difficult that you've set your mind to?'

I don't think I've ever solved the classical problem of the interface between science and politics, though I've given a lot of attention to it. I think it's a peculiarly hard interface. I had to work hard at it as director of RAE, and I got on very well with some of my Ministers of Aviation, particularly Julian Amery, but I always felt that there was such an enormous barrier between the way in which politicians look at the world, and the way in which scientists do, that it's very difficult to penetrate it. And yet it must be penetrated. The interests of science demand that scientists and politicians have to find a way of coming together and talking together, so I think there is a very hard problem there, and I know that very few scientists have successfully penetrated that barrier, although so many have attempted it.

'Do you think it's just a totally different mode of thought?'

That is the main problem undoubtedly, because the scientists look at things in one way—they're obsessed with truth and accuracy, and also concerned with the long processes in time by which truth is discovered. Politicians want answers very quickly, and they're more concerned with how things are presented to the electorate, if you like. Of course, they have a lot of right on their side. I have to be a politician in the college to carry everyone with me as far as possible, and they have to carry a whole nation with them, so their priorities are necessarily different for very good reasons, but nevertheless we do have difficulty getting on to the same wavelength.

'Do you think this problem with politics is peculiar to science, or do you think it would be true of almost any non-political activity?'

I was using science to mean, you know, the science-based disciplines, including medicine and engineering and so on. I don't think the other side of academia, the arts side, has such a difficulty actually.

'Why do you think that is?'

Well, I think they are very much concerned with communication and so on, and, of course, it may be partly that a number of politicians, probably the majority, have an arts background and, furthermore, studied the arts at a time when the arts and the sciences were very separated from each other.

'Can I ask one last question? Is there anything in the academic world that doesn't interest you?'

[Laughter] I've only really tried to study about sixty academic disciplines, and only the ones that are studied here at University College, and I must say I've found them all absolutely fascinating. I really haven't got any reservations on that; so I cannot believe that if I did try to study some of the other subjects—agriculture or something—that I wouldn't find them very interesting too. We don't cover everything by any means here, but certainly I've found that being the head of a great academic institution is a marvellous way of getting you interested in almost every academic field.

AGAINST THE GRAIN

JAMES LOVELOCK
*was born in 1919 and is an independent scientist living
in Cornwall.*

CHAPTER 8

Freeing the mind

❧

James Lovelock
Chemist

THE idea of a freelance scientist working from home sounds absurd. A well-funded laboratory, colleagues, tenure, are all regarded by most scientists as not only desirable but essential. This is the time when big is beautiful. How then has James Lovelock managed not only to be independent for the past twenty-five years, but also very successful?

Lovelock's formal training was as a chemist, and until 1961 he worked at the National Institute for Medical Research in London. Then he left, and has since financed himself through inventions and consultancies. This has allowed him to pursue the research project for which he is best known, the quest for Gaia. His thesis is that all components of our planet, both physical and biological, can be regarded as if it were a single living system, a self-regulating being, compensating within itself for change or damage so as to keep the whole in equilibrium. At first sight this seems a rather romantic idea, not at all like hard science, and when first put forward it was regarded with deep suspicion. But now, twenty years later, it has been taken increasingly seriously. Recently it was the subject of a prestigious international conference in the United States. It is a remarkable achievement to have developed a major theory about the world, quite outside conventional science. But if Gaia and Jim Lovelock's way of doing research is unorthodox, his inventions, which have been enormously successful, place him firmly in the pantheon of orthodox science, and he was elected a Fellow of the Royal Society in 1974.

His main contribution has been the development of detectors which have greatly extended the technique of gas chromatography, a way of separating out and identifying chemical compounds. The most important was the electron capture detector, a most sensitive device for detecting the presence of minute quantities of chemicals. It has a very wide range of applications, but Lovelock himself has used it primarily to study the earth's atmosphere. It was his measurements of chlorofluorocarbons that led to the recognition of their effect on the ozone layer. The use of the detector to study pesticides in the environment, provided the background information which Rachel Carson used in her book, *Silent Spring* (1963), and so started the environmental revolution.

Jim Lovelock lives with his family in a modest house in the Cornish countryside. His office is also one of his workshops, and we sat down to talk surrounded by the paraphernalia of electronic apparatus and computers. The instructions to find the house had been detailed, but we had arrived late after a somewhat anxious drive through tortuous and seemingly endless country lanes. Lovelock too had only just returned, hours late, and after various delays, from London. My first question had to be, was the self-imposed isolation essential?

—————

More or less, yes. We are hermits by nature. We're not anti-social, but we just like a quiet life, and the way science has gone for me in the last ten years, is that the numbers of letters and visits and phone calls that come have really got quite uncontrollable. If I had a large department to handle it, it would be all right, but there's no-one else to do it. So we protect ourselves. Our address is unlisted, if you like.

'But that's very contrary to how one sees science, which is as a social activity. You are really very isolated here.'

Well there's room for both. I mean in the arts you have the cathedral builders, and the orchestras. These are all very social activities. But you also have the individual artist painting his picture away in the garret, and the novelist working in a country village somewhere, quietly, not wanting loads of people. The same should be true of science, but for some reason the whole subject has gone completely wrong.

'How do you actually function?'

That's a good question. I don't know, to be quite honest. I'd better explain how I became independent first of all because it all follows on. I worked for twenty years for the Medical Research Council, and they were almost the perfect employers, they couldn't have been kinder, more considerate. I was astonishingly well paid, and given complete freedom to do what I liked, but the one problem was tenure. Not that I didn't have it, but because I did! It made me feel that there were tramlines of inevitability going on all the way down to retirement and the grave. And this was a most stifling thing. It was dreadful. I knew that nothing I did—short of something criminal, or seducing the director's daughter on the front lawn or something like that—would get me dismissed. And that even if I did no work for a couple of years, I was secure. This was an awful feeling, and I think any artist, or novelist would understand it. It's not the sort of thing that's good for creativity. So I decided the only thing to do was to go independent.

'You really rejected it so strongly? Most scientists feel very strongly about tenure, and want it.'

Yes, and I think that they may well be mistaken. A possible explanation is that people who are nowadays called scientists are not really scientists, any more than advertising copywriters are literary people. They may be able to write beautiful copy, but it's not quite the same thing. They are in a job, a career. A scientist shouldn't be, I think. A scientist is much more like a creative artist, somebody who does it for a vocation. It's the only way of life they want, and nothing else. They don't really think about where next week's money's coming from—at least, they shouldn't.

'This is a very romantic view of science.'

Very romantic, but it's worked for twenty-five years. We've never known where the next income was coming from, and we started off with no funds whatever and four children to raise. It's never failed.

'So then how do you actually live? You say that you've never known where the money's going to come from, and here you are, in isolation, how do you actually live?'

Well, I live mainly by inventions. I happen to have, how can you call it, a talent for solving technical problems quite well. And in the same way that an artist makes pot boilers to pay for the work he wants to do, so I have to make pot boilers in the way of inventions that fund my science. The strange thing is, it all feeds back on itself. You see I found in science that the only things worth doing are interesting things. If you only do interesting things, they always lead to new inventions and things. You start asking questions for which there are no instruments with which to make the measurements that'll help you find the answer. So you have to invent the instruments. Then other people want them for other purposes, and will buy them from you. But it's solved your problem. As it happens, invention is quite easy. Most people are inventors if they only knew it. What is really clever is thinking of the need.

'But when you invent, how do you actually get your ideas for inventing?'

Well first of all I've got to get the need, we've agreed on that. What I tend to do is to wake about five in the morning—this happens quite often—think about the invention, and then image it in my mind in 3D, as a kind of construct. Then I do experiments with the image.

'In your mind?'

Yes. Sort of rotate it, and say, 'well what'll happen if one does this?' And by the time I get up for breakfast I can usually go to the bench and make a string and sealing wax model that works straight off, because I've done most of the experiments already. That's the process I use.

'What's your pot boiler though?'

Pot boiling for me is what the family refers to as gas pornography, or in proper terms, gas chromatography.

'So your skill, you're saying, is in measuring things?'

That's right.

'That doesn't seem to be the interesting part of science, which is about understanding things.'

Precisely, but the two are tied together. You can't understand things unless you have the numbers that your measurements will give.

'What sort of things do you measure then?'

Oh, minute traces of interesting gases in the air. Can I give you an example? In 1972 I was interested in how the world regulates itself. One of the interesting elements was the element sulphur and I had a notion that the algae that live in the oceans produced a gas called dimethylsulphide that went from the ocean into the atmosphere and transferred sulphur from the sea back to the land, and made up the missing amount of sulphur, without which the vegetation would die. Now it was very difficult to get support to do the work, in fact, it was impossible because at that time it was contrary to normal scientific teaching. But with the help of some kindly civil servants in the Natural Environment Research Council, I took a ride on one of their ships, the *Shackleton*, from Britain to Antarctica and back, and measured this compound in the oceans and in the air throughout the course of the voyage, and found that it was distributed globally. It's now recognized that it is indeed the major component of the natural sulphur cycle, and a very important one. We're finding there wouldn't be any clouds over the ocean were it not for the production of this gas, and without the clouds, the earth would be very much hotter than it is now, and so on.

'But couldn't you do this work within conventional science?'

No, because you couldn't get funds to do it. I put in for grants to make the measurements, and the reply came back, not from the civil servants, but from the peer review committees of academics, 'oh this is nonsense, everybody knows that the sulphur cycle is completed by the emission of hydrogen sulphide from the oceans.' Now I've been on ships enough to know that hydrogen sulphide doesn't come out of the water, it stinks, I mean you would have smelt it, but academics, who live in universities, and never go anywhere, don't know.

'Are you slightly hostile to the way science is organized?'

Yes, very hostile. It's become a career business nowadays. You see, even when I was young, pre-World War II, science was very much a vocation, and I remember the headmaster of my school telling me, when I said I wanted to do science, 'Oh you're a fool, Lovelock. The only people that can do science are those with private means or genius, and since you have neither, drop the idea at once. You may be fascinated now, but when you come to marry and raise a family, you'll bitterly regret your decision.'

'What did you actually do then?'

All I wanted to do was to work in a lab so I didn't care. I took a job as a laboratory technician for a firm of consultants in London. Fortune smiled on me because they were a remarkable firm. They were people that would tackle any problem, regardless, and they treated us youngsters who went to work for them by throwing us off at the deep end, and expecting us to do everything, learning the hard way. And by the time I'd spent a year and a half with them I was a pro, scientifically. If I did measurements, they had to be right. They weren't just like an average student is taught, 'oh well it doesn't matter, as long as you understand the method.' People's jobs depend on getting the answers right.

'But where did you get your formal academic training?'

Well, they paid for me to go part time to Birkbeck College, which is part of the University of London. That was one of their decent things. But after I'd been at Birkbeck a year, World War II broke out, and on the results of my performance at Birkbeck, I was able to get a small scholarship to go to Manchester University to complete my degree, which I did. And I must admit I found university excessively dull after the particular life I'd been leading, and learnt almost nothing there.

'But you chose chemistry didn't you?'

Well I knew you had to have a union ticket in order to be allowed to do science, and it didn't matter very much what it was. I would have rather have done physics, but my maths is a bit slow, and I thought, well chemistry was the easier option, so I chose chemistry.

'Did you go on to do research then? To do a Ph.D.?'

No I didn't. I would have liked to have done, but we were not wealthy, and I couldn't afford it. Again, fortune smiled, because I was sent by the professor of the department for a job at the National Institute for Medical Research, which then was in Hampstead, on Holly Hill, and, of course, I couldn't have been sent to a better place as an apprenticeship for a young scientist. It was an institute that was stuffed full of really brilliant scientists, and they were all on *ad hoc* war research problems. During fire-watching I used to have to share the roof with distinguished old boys, Nobel prize winners and the like, and when the bombs were dropping, or the doodle bugs coming, it caused them to have a brain dump, they felt they had to pass on all of the tips, all of the information they'd gathered in a lifetime, to this youngster in his early twenties standing there. It was wonderful, it was the most amazing apprenticeship anybody could ever get.

'You never got a Ph.D.?'

Yes.

'You did get one eventually?'

Yes, in medicine. It was the rule of the game, not during World War II, but afterwards, that any research work you did at an institute, like the National Institute, could qualify for a Ph.D. degree if you had an appropriate supervisor, which was one of the staff members of the institute. So I did it that way, and took a Ph.D. on 'The Transfer of Infectious Diseases' which, of course, is medicine. I see no barriers in science. I think that it's one subject. The territories, the disciplines, are purely feudal, set up by professors to retain territories over which they have control. I think there's no sense in it at all.

'So how have you managed to cross these boundaries? There are these feudal boundaries, you're right.'

I accidentally found I had a passport, and my passport was hardware, particularly this electron capture device, because it's the most sensitive analytical device there is in existence. It can measure the less than nothing of something, so to speak. And it's amazing the numbers of people in every discipline that want to be able to do this. I mean, physicists such as meteorologists, for example, want to measure the movement of air masses, and it's quite easy to label the air mass with a small amount of an inert, harmless, detectable compound that can be measured this way, so they come to me and say, could I make them an apparatus to do this, and then they start talking about their problems to me, and I go right across the barrier.

'So that invention gave you your passport?'

Yes.

'How did you come to that invention? Did you see a need, as you mentioned earlier?'

It was utterly messy. What happened was, I was working on the freezing of animals alive, and then bringing them back to life. Something I wouldn't do now—I've got a bit soft in my old age—but I was a youngster in those days at Mill Hill. The brief I had was: what was the nature of the damage to living cells by freezing? I soon found that it was the problem of the membrane constituents, which are fatty acids, lipids and things like that. And I wanted to find out what was the difference between the ones that were resistant, and the ones that were easily damaged. So I needed to analyse them, and Archer Martin had just invented the gas chromatograph, so I took a sample along to him, and said, 'Could you analyse the fatty acids there are in this because I want to know?' He said, 'Oh sure we can, it's just what the gas chromatograph is for. Give me the sample.' I handed it to him.

'Well, where is it?'

'It's at the bottom of the tube.'

'Well how much is it?'

'Just a few micrograms.'

'Well,' he said, 'that's hopeless, we'd need milligrammes before we can do anything.'

So I asked him what I could do and he said I'd just have to go away and invent a more sensitive detector, ha, ha. So I did. I knew that ionization devices, which are very physical, represent an exquisitely sensitive way of getting a signal from a chemical composition in a gas mixture, so I thought I'd play around with radioactive sources, and gases and things, and see if I couldn't invent some ionization detectors, which I did. Now one of them was a wildly unpredictable and incredibly sensitive, but almost useless, device, which was so sensitive that if you put a little bit of something in it, you couldn't use it for a week. And this was the electron capture detector. Nobody wanted it, of course, it was far too messy. But I invented another one called an argon detector which really was very useful, and solved a lot of the problems, including the fatty acid one I've mentioned. But I kept going back to this electron capture detector because I was curious about it, and gradually developed it to a point where it did work, and it turned out that it's exquisitely sensitive, almost—though not entirely-—exclusively, to nasty things, the things that are carcinogenic, or environmental hazards and, of course, it paved the way for Rachel Carson's book, *Silent Spring*. Without the data that that device had gathered she would never have been able to write her book. It was what found the pesticides in penguin fat, and mother's milk, and all those kinds of things.

'But you didn't do the measurements yourself?'

No, others did them.

'But was it her book, *Silent Spring*, that got you interested in the whole concept of what ultimately led to Gaia?'

No, no it wasn't. It was space research did that. After I'd invented these ionization detectors I was entering this phase when I was getting unhappy about working at Mill Hill, because nice as it was, there was this tenure problem that kept bothering me, and I daren't leave because they were so kind, and such nice employers, that I just didn't have the heart to say to them, 'I can't stand this.' And then out of the blue came this gorgeous letter from the director of Space Flight Operations at NASA. In those days NASA was only about a year old, I think, and it was only three years after Sputnik had gone up, and this letter was saying, would I join in as an experimenter on the lunar soft landing mission, using my apparatus to measure the lunar soil. Well, I've read science fiction since I was a kid, and it was an offer that was, you know, the sort you couldn't refuse. So that got me off the hook.

'You went to America?'

I went to America, not intending to stay, but to participate in these NASA space activities. And it was all done in a rather nice amateurish way, a bit like World War II science. But they kept talking about detecting life on Mars, and this made me very curious because I'd worked a lot in biology as well as other sciences, and I began to find that the experiments that they were proposing to send to Mars were, to say the least of it, asinine. What they were planning was to send things to find Earth-type organisms in the Martian environment, and this seemed to me absurd. I mean, if you landed a spacecraft in the middle of the Antarctic, you wouldn't find any life. Why should you expect to find it in the middle of an infinitely more hostile Martian desert? And I felt this was the wrong way to go about it. And at the same time I read Schrödinger's little book *What is Life*, as I thought this was apposite, and it seemed to me the only way to do it was to measure the composition of the planet's atmosphere—if it was a dead planet, then all of the gases would be in equilibrium, but if it was a live planet, the organisms would be obliged to use the atmosphere as a transport medium for their raw materials, and their waste products, and this would change it chemically in a way that would make it recognizably different from a dead planet. This fed right back to my expertise, and that's how I got into life detection, and ultimately to Gaia. Because when we got the results from Mars and Venus—not by spacecraft actually, but from infra-red telescopes—it showed that both planets were close to equilibrium, and therefore, according to this idea, dead, but that the Earth was wildly out of equilibrium, incredibly out of equilibrium. Our atmosphere is almost like the gas mixture that goes into the intake manifold of a car, a mixture of oxygen and hydrocarbons. If it were more concentrated it would be combustible, or explosive, and for such an atmosphere to stay in a steady state for hundreds of millions of years is an impossibility, unless something's regulating it.

'Was your idea about Gaia well received when you first proposed it?'

Oh no. It's only just beginning to get received, and that's getting on for twenty years later.

'What was the hostility towards it?'

Mainly from the biologists—they're the most vociferous—because, I think, they have been crippled by their battle with the fundamentalist religious people. What's happened is, the biologists are being forced to argue in fundamentalist terms. Without ever realizing it, they're beginning to use the same language as the creationists, and they're beginning to treat Darwinism as if it were a received text and dogma, and anything that appears to be, even on the face of it, slightly contrary to Darwin, must be wrong. You don't argue about it, it's wrong. And Gaia seems to be contrary to Darwin. It isn't. It's absolutely in accord with what Darwin said, and I'm sure if Darwin were alive now he would find no difficulty in the concept. But not his disciples.

'Have you tried to promote it, the idea? I mean have you tried to fight the biologist world?'

So far I haven't, but I'm just about to.

'Why are you prepared to do that now?'

Because I'm very much more certain about how Gaia works, and I understand it very much better, so I feel in a strong position to answer their criticisms.

'Has your information accumulated then?'

It's not only information, it's models mainly, numeric models on computers that demonstrate how a Gaia-type system can, in fact, work. You see one of the main biological criticisms was, there is just no way that Darwinian natural selection could lead to altruism, global in scale, it just couldn't happen. And what I've been able to demonstrate is that that criticism is pure dogma. What the biologists left out was that organisms do not just adapt to a given environment. That environment is determined by the rules of physics and chemistry—which they would like to know about, but that's not in their department, that's something that happens across the quadrangle of the university. In fact, the environment to which the organisms adapt is, as I've put it slightly poetically, the breath and the bones and the blood of their ancestors. And they themselves are changing it, so it's a closed system entirely, and under such circumstances, any organism that changes the environment in a manner that will allow it to leave more progeny, will succeed. Whereas one that changes the environment in the way that will adversely affect its progeny, is doomed, and that is how the system evolves. And I don't think Darwin would have disagreed for a moment with that.

'In spite of your hostility, your avoiding conventional science, you are a visiting professor at Reading University. This seems like a backsliding to me in your terms.'

Well I suppose it is in a way. You can't ignore the system altogether, and it isn't a bad system really, it just doesn't happen to suit me. And what happened early on after I became an independent was that I was shocked to find that people who had published my papers before, like the journal *Nature*, had published them not because I was me, but because I was working at Mill Hill. And when I sent in papers from my home address they bounced straight back again, and one of the editors told me, 'oh we never publish papers that come from home addresses, they always come from cranks.' And I was telling a friend of mine, Peter Fellgett, who was professor of Cybernetics at Reading at the time, and he said, 'well why don't you become a visiting professor here, it would solve that problem for you.' And I said, 'oh well, thank you very much, I will'—and I still am. And it's a very interesting relationship. No money's changed hands. The only thing that's really ever happened is that I give an occasional seminar there, and a graduate student came to work for me from Reading, who turned out to be one of the best students I've ever been associated with.

'But has the scientific establishment been supportive of you? Apart from the particular opposition to Gaia, have they been sympathetic to your independence?'

No, not at all. I've received virtually zero support from the British scientific establishment.

'Why do you think that is?'

Well I haven't asked for it, that's one reason. You can't expect them to, sort of, come out of the blue and do it. I stopped asking after the debacle with the dimethylsulphide and another grant application which coincided with it, to measure the chlorofluorocarbons in the atmosphere. This was also turned down, on the specious grounds by academics, that it would be impossible to measure them therefore the proposal was bogus. So, and as you well know, I went out and measured them on the same voyage, and that was what started the ozone chlorofluorocarbon controversy. So one could say it was a very worthwhile piece of work that was done on home funding.

'Your ideas, at least about Gaia, and lots of your others, are very synthetic, you bring together things from many different fields. Do you think living in the country has had some importance?'

Oh yes, I think you're absolutely right. I think one of the problems nowadays is that ninety-five per cent of us live in cities. It's an enormous proportion of the human population, and as a result we suffer, those of us that live in cities, from sensory deprivation. I mean, just think, how many people do you know that have seen the Milky Way recently at night? It's a gorgeous sight if you go out in air clear enough, as you can here on a cloudless night, and see our Galaxy stretching right across the sky from horizon to horizon. You cannot see it in any city anywhere because of the lights. And how many people have lain on their back in a meadow and smelt the sweet fresh air, and heard the birds singing? You just can't do it any more, a city park is not quite the same thing.

'But how does that help your science?'

Well I think if you can, you feel part of the world, you feel much more interested in it, and your sense of wonder is stimulated.

'Is there anything in your background that you can see that has made you the sort of person you are, which is a rather unusual scientist?'

The only thing I can really pin down there is going back to what I said earlier, which is the apprenticeship type of training I had as a youngster, of being thrown in at the deep end to deal with a different problem every day almost, and an *ad hoc* one at that. It made me have to be broad, right across all of the fields of science, and as a consequence I never felt that science was a divided set of disciplines. I felt it was just one interesting, gorgeous subject, well worth spending a life working on.

THE LATE PETER MITCHELL (1920–92)
*was Director of the Glynn Research Foundation,
Cornwall.*

CHAPTER 9
Gardens of the mind

∽⟩∼

Peter Mitchell
Biochemist

At the peak of his scientific career in the 1960s, the late Peter Mitchell spent two years not doing science, but working full time on the restoration of Glynn House, a large and beautiful Regency mansion in Cornwall. Only a year or so earlier, while director of the Clinical Biology Unit at Edinburgh University, he developed a theory which was to lead to him being awarded the Nobel prize for chemistry in 1978. His time as architect and master of works turned out to be only a temporary retirement from active research. But his departure from the formal scientific establishment was permanent. He built a personal laboratory in Glynn House, and set up the Glynn Research Foundation, a registered charity, to promote fundamental biological research for the benefit of humanity. It challenges many of our ideas about science that a distinguished scientist should abandon his research at a critical stage, and then prefer to continue outside the conventional framework, and in apparent isolation.

Mitchell's Nobel prize was for his work on the chemical processes involved in the generation and transfer of energy within living cells. No scientific idea is completely new, but some are so novel that when they are first put forward it takes a long time for them to be treated seriously. That was the initial fate of Peter Mitchell's idea that in living cells there is a mechanism for power transmission, analogous to power transmission by electricity. At the time, biochemists thought that there had to be chemical intermediaries transferring the energy captured from sunlight in plants, or the oxidation of carbohydrates in animals, into the substance ATP where it is actually stored. Mitchell proposed that the energy was transferred by the flow of protons, positively charged particles, and that the intermediate substances people were looking for just didn't exist.

Peter Mitchell's approach to all scientific problems, and to life in general, was always original, humane and positive. In the relaxed and romantic setting of a lovely country house, with his white hair, gentle face, and quiet way of talking, he gave an impression rather of spiritual than scientific wisdom. Given these apparent contradictions, I wondered why he had chosen to become a scientist.

———

I suppose I was pushed in that direction when I was very young because I tended to be better at thinking about the general rather than the special. I always had a bad memory for special things. For my own part, I suppose I became a scientist because I was interested in the relationships between things in the world. I was never a sort of 'stinks' man.

'But you went into chemistry?'

That's a funny way of putting it. I don't think I did really 'go into' chemistry. I remained a human being. But to answer your question, I got interested in engineering first, actually. When I was a small boy I loved making things. I was lucky enough to have an elder brother who actually became an engineer, and he'd got a bit of a workshop. But I tended to be the one who wanted to be more active, and I was always making little engines and things. I suppose that helped my development as a thinking person, because of the relationships between shapes. That's something I've kept in chemistry, of course, because a lot of chemistry is really a sort of physics, and so you're concerned with the relationships in space of the atoms.

'But I still don't understand how you chose chemistry, or why?'

You say I chose. I don't think I've ever really *chosen* anything much in my life. Sometimes, of course, you make a deliberate move towards something because you recognize that you have, I don't know, a 'capability' in that direction. Perhaps the word 'capability' is too conceited. You recognize that there are *possibilities* there for your psyche, for your soul, for your being to fulfil itself. I think that's what made me do what I have done, but I think I might just as well have become a farmer as a 'chemist', as you call me; though I don't know that I really am.

'But you did choose to become an independent scientist. That was a very clear choice.'

I have become very independent in the way I have lived, but I don't think I deliberately chose it. Originally I worked in Edinburgh, where I was deeply interested in the bit of teaching that I had to do, and also in the small research group that I was running. But because I found difficulty in ordering my mind about—I've never liked dictatorship either outside myself, or within myself— I got acute gastric ulcers, and had to leave. And that was really how it came about that I became more independent. It wasn't a very conscious decision: I became ill because my position wasn't appropriate for my personality; that drove me out of Edinburgh, and by an extraordinary set of accidents, it drove me in to Glynn, where I founded this present institute with my colleague, Jennifer Moyle, in 1964.

'Did you not like the academic life?'

Oh, I loved the academic life. I still love the academic life, but I suppose one of my troubles has been that I've always been interested in so many different things. So in a university environment I tended to become interested in everybody else's problem, as well as my own problem, and that meant that my energies were dissipated.

'When you made the move to come down here, you actually did give up science for a time.'

Yes. What happened was that while I was still at Edinburgh, I had come into possession of a mansion called Glynn House, which was in a state of ruin. Old buildings were a hobby of mine and I'd originally hoped to do something to prevent the continued ravages of the dry rot and so on, and get it back on the market. But when I got ill and had to leave, I went down to live in a cottage next door to Glynn, and started to think, well, what am I going to do now, because I can't be idle. So the possibility arose that one might found an independent institute. As it was most unlikely that you could do that all on your own, I wrote to Dr Moyle, and said 'I have a mad idea. Would you be prepared to give up science for two years and work with me to restore Glynn House, and then try and establish an independent charitable foundation?' Of course, she should have written back and said 'No', but she didn't. She wrote back and said 'Yes'. So for the first two years, Jennifer Moyle and I did no research but worked out of a little office in Glynn House with a gang of twenty men and a foreman. We had very little money and we did all the quantity surveying ourselves, and all the drawings and arrangements with the planning authority, because it was a Grade II listed building. And only after that could the scientific work begin.

The institute shouldn't have succeeded, should it, because it was 40 miles from the nearest university, but funnily enough it does seem to have worked.

'As you say, your experience of having a small, independent, research group out here in the beautiful countryside is absolutely contrary to all the current ideas about how one runs science, and how to do good science—that you must have large, centralized groups, with lots of people interacting.'

I agree it's against the traditional notion but, personally, I don't understand why the traditional view is what it is. One of the things that happened when we started work out here at Glynn was that our communications with the rest of the world greatly improved. The reason for that was quite simple. There is a difference between noise—by which I mean the transmission of activity between different centres—a difference between noise, which is information that most of the time you don't need, and actual information transfer. Also, the current notion is that scientists exchange information at big conferences, but I think in one's own experience, and I'm sure many scientists would agree, that in the bustle of the big conference where the important scientists are giving great papers and things, the bustle doesn't allow any real communication between the minds that are operating at the frontiers. You don't have time. Also it's a little difficult for the right people to meet each other because there are so many others around. Whereas in a centre like this, if somebody visits, it takes them four hours to get here from London, so they're likely to stay. Helen, my wife, is a very good cook, and we like good wine and things, so you have a sort of captive situation in which you probably do communicate with one another.

'You've obviously enjoyed being down here, and it's been very successful for you. Have there been no disadvantages?'

I don't think there have been substantial disadvantages. Lately, of course, we've had problems with funding. But I see that as an advantage as well as a disadvantage, because I myself have now begun to spend a large amount of my time trying to communicate better with the public. It's my feeling that the public doesn't understand very much about the real process of scientific investigation, so probably it's rather good that some of us, in order to get our bread and butter, have got to convey to the public, in a much better way, that science is really a wonderful cultural activity and belongs to the countryside as much as to anything else. I mean, how many people realize that science itself, or the ongoing process of science, is really the gentle art of investigation into the nature of our world and of ourselves. It *is* a gentle art, and there's no reason at all why it can't be practised in a house in the countryside.

'Did you set out to solve a particular problem as a scientist?'

I said earlier that I tended to be interested in the general, in the relationships between things. That being so, the broader the problem you manage to solve, the happier you're likely to be. So I think the answer to your question is that there was one particular problem which my background really prepared me to look at, and that was the problem of the relationship between the chemistry that goes on inside a living cell—which is basically the transformation of one thing into another, A is turning into B—and the real action that's going on in a cell, or a group of cells, or in a whole organism when it runs about, thinks, looks, or pays attention. So the question in my mind, a very big question was, how do the two get connected? And at an early stage in my career as a biochemist, I began to think that it was really a sort of a non-problem. The orthodox way of looking at it was to ask how is metabolism, as it was called—that's the chemistry changes in a living cell—how is metabolism coupled to transport, the movement of things through membranes or along fibres? How is it coupled? And I came to think the answer is that it isn't coupled, it's one and the same thing. Nature itself doesn't have physiologists, on the one hand, who look at the action, and biochemists, on the other, who look at the chemistry. Nature itself does it as one thing. So what you need to do is to try to bring together the notions of chemistry and the notions of physiology, and basically that's what I've tried to do and to some extent it's been successful.

'Well it's been very successful. But you're really saying that it was from a philosophical point of view that you came to your really seminal idea?'

Oh yes. You've hit the nail right on the head there. When I was first at university I became very interested in the Greek philosopher, Heraclitus, and I began to think that there were two kinds of things in the world. One kind, which I called for myself 'statids', were like teacups, which don't evolve in any way except that they gradually get broken, they get less recognizable. And the other sort were things like rivers, which Heraclitus talked about, which flow. The identity of the river is different, it's environment is constantly going through it. And so I called these things 'fluctids', 'flow things'. Flames are like that, and people are like that.

'But, Peter, are you really telling me that your highly technical theory of how you actually get energy made in the cell was really based on these very general, and somewhat romantic images of the world?'

Yes, completely so. And I especially agree with the word 'romantic'. I think this is something which more scientists ought to explain—that we don't do science because we're scientists, because of science—we do it because we're human beings. It is a most wonderful romantic, cultural activity, just as much as being a sculptor, in fact, more so because, of course, everybody can be doing it all the time. It's problem solving; what you're doing from when you first get up in the morning.

'The actual idea, OK, it came from these philosophical ideas, but there were the technical details to be worked out. Did you come at them all at once, or was it a slow process?'

Oh it was a very slow and painful process. Most people who try to be creative, I think, have found that they've got to become craftspeople as well as art people. You have to go through the, dare I say, dreary business of school, and in my case, of learning chemistry out of the textbook. There is a huge amount of information that you have to absorb before you can start walking about in it, as it were. I've often thought that the human mind is a bit like a garden. You prepare this garden, and you plant things there, and it's a sort of garden partly of facts, and partly of ideas, and you keep re-arranging it, and that's really quite hard work, but at the same time, it's well worth it because you can go for a lot of walks, especially if you don't sleep very well.

'When you got the answer, did you know that this, as it were, was right?'

No, I don't think you ever know, because it never is right, is it? When you're trying to appreciate the nature of the world, you're looking at the real world, and you're making images of it as if you were a sculptor. The images are in your mind, and you transfer them to flat pieces of paper. You think you've got a good model, but it's still only a model of the reality, isn't it? And so, when you think you've hit on an answer, it's never going to be more than a partial answer and that's very good for your modesty. And then, of course, you share it with your scientific colleagues. That's another aspect which is wonderful, because whether you turn out to be wrong, or right, everybody wins. This is something else I think that even quite a number of scientists don't seem to appreciate, that being wrong in science is often much more fun than being right, because the next day you wake up with a new horizon, with a new set of priorities for the next attack you're going to make.

'That's a very nice feeling that you can accept being wrong. I mean we all have to live with it, but it's not something that most of us like. When you put forward your rather radical new idea, in 1961 how did the other scientists take it?'

At first they didn't take any notice. It wasn't an entirely novel suggestion, of course. That's another very important thing to say, that it's almost impossible to have a completely new idea. When you look back in the literature, it had been suggested before, though not in a very solid way. What I did was to propose it as a proper hypothesis which would be falsifiable. But nothing really happened in the way of experiments that were designed to test it until about 1964, when Jagendorf in America started to look at chloroplasts, which are the green particles in leaves of plants which absorb sunlight. He found that when he shone a light on isolated chloroplasts, protons started whizzing about, as the theory had predicted they should, and he was greatly excited by this. So that kicked off some experimental work, but it had taken three years from the publication of the original paper. It took about twenty years altogether. But it did turn out, very gradually, that the hypothesis seemed to be more or less right.

'Why did the rest of the scientific community really just ignore you?'

Well they had a very good reason to ignore us I think. There were precedents in biochemistry for thinking that there ought to have been chemical intermediates between the two kinds of process. But biology has been very adventurous through evolution, and all sorts of new methods have been discovered for doing basically the same thing. It was, however, very unexpected that in this case a quite different sort of system would operate.

But there was another very big difficulty in that most of the people who were interested in this field were chemists. They were the sort of people who were experts at studying the chemical intermediates. They weren't really expert physicists and, as it were, my work said they were stuck working on the wrong thing, and that didn't please them. And you can imagine, if you were working on something, thinking you were going to solve a problem, and then somebody told you that what you're looking for didn't exist at all, you might not be at all pleased with that. So I got a good deal of criticism, and a certain amount of crotchetiness, let's say, from my colleagues.

'Did you mind that?'

Did I mind? Well of course I minded. I minded very much because I respect my colleagues, and anyway I thought they might well be right. I think you're bound to mind, you're a human being. I suppose if you happen to be very callous you wouldn't care, but most of the scientists I know are rather gentle people who mind about criticism. But they wouldn't necessarily alter their view, of course. That's a different matter.

'Did you ever think of altering yours?'

Did I ever think of it? Well I certainly never intended to alter it, at least, not unless or until it was proved to be wrong. But I think we ought to come to a more recent time, because although the general theory that I proposed in 1961 did turn to be right, in certain details, I was wrong. And so you have to see it both ways, don't you? These arguments, you regret, of course, that sometimes they become a bit personal, but again I think this is what's so wonderful about science, that everybody wins.

'It's true that science has gained, but when you were wrong after all that time, in a way you did lose.'

Emotionally, of course, it makes a considerable impact on your personality.

'It makes one depressed.'

Well it might make some people. It didn't make me depressed, it made me, I must admit, it made me sort of excited. I was rather pleased that I was wrong because it allowed a resolution. It does make a personal impact, of course, but this is something you need to bear because there are much more important things than the particular dent in your own personality.

'Perhaps you can only deal with being wrong because you were so right that you got a Nobel prize. Did you ever expect to get that Nobel prize?'

Oh no. No, I certainly didn't expect it.

'And the consequences?'

To some extent, I think, having become a Nobel prize winner tends to put a certain distance between you and your colleagues, and this is something which I've regretted. Of course, 'winning', I've just thought, is not a good word really, because the fact that such a prize comes to you is hardly of your doing. I was especially grateful because the colleagues who presumably worked quite hard to vote that prize in my direction, had initially totally disagreed with the hypothesis that I put up. I was enormously heartened by that, and my feeling about human nature was tremendously, sort of, boosted. You love your fellow human beings even more.

'Now a major portion of your scientific life has been collaborating with Jenny Moyle.'

Yes.

'What's the nature of that collaboration?'

The nature of that collaboration was that Jennifer was an exceedingly good experimentalist, and so very often when I was busy writing reviews, or trying to develop new ideas to carry the work forward, Jennifer would be vigorously pursuing the experimental work.

'So many of the experiments were actually physically carried out by her rather than yourself?'

Often, of course, we were both at the bench together. But often as well, Jennifer was at the bench getting on with the experiments while I was sitting about thinking.

'Do you think she was happy in that role?'

I think she was very happy in that role. But then you see she is a very versatile sort of person. Since she retired several years ago she's been having a very full artistic life. And maybe that's been something about both Jennifer and me. I've often thought we're not really scientists at all, but we're just people, and we happen to have spent a lot of time doing science.

JOHN CAIRNS
*was born in 1922 and was Professor of Cancer Research
at Harvard University. He now lives near Oxford.*

CHAPTER 10

Not a company man

⤳⤳⤳

John Cairns
Molecular biologist

THE overwhelming impression from the scientific literature is that researchers are an orderly lot who spend their time working in groups and turning out innumerable publications. John Cairns is not like that at all. He is a brilliant and original experimentalist whose career has been shaped as much by accident as design. Working mainly alone, and publishing relatively little, he has tacked across continents and disciplines, challenging orthodoxy as he goes. Some of his most recent experiments, which he undertook in the context of cancer research, have upset the evolutionary establishment.

Current theory claims that all mutations, the variations in cells or whole organisms on which natural selection acts, arise purely by chance. What Cairns's experiments have shown is that under certain circumstances the mutations which appear in bacteria seem not to be random but directed towards helping the cell survive.

Cairns trained originally as a doctor, and I wondered whether he had started out with any thought of where it might lead.

———

I did nothing very intellectual. I went to medical school, mainly because my father was a doctor. That's the kind of thing that happens. And in medical school I came to realize my father was so distinguished that I had to get out of medicine. So, after doing a few house jobs and things like that, I drifted into clinical pathology, and from there into viruses and from there to molecular biology. So, I think my career was one of escape all the time.

'Why did you have to escape?'

Because I felt that I could never succeed and feel that it was by my own work rather than the beneficence of his shadow.

'You say you drifted into biology. There must have been something that . . . ?'

Pushed me into that?

'Yes.'

Well, as a clinical pathologist I'd done things like post-mortems and haematology but the bit that I liked best was the bacteriology.

'What was nice about it?'

I don't know. Maybe I feel a certain affinity with bacteria in a colony on a plate, or maybe it's the smell of bacteria growing on plates. It is a rather nice smell. When you come into the laboratory in the morning there's this homely smell greeting you, and so I liked bacteriology. Then I had the opportunity to go and learn about viruses in Australia in McFarlane Burnet's lab in 1950 and that, too, was an accident. The Rockefeller Foundation had run laboratories all over the world to tackle the problem of yellow fever, and eventually developed a vaccine against yellow fever, which is the best vaccine that has ever been produced against anything, since one shot protects you for life. Once they realized they had solved yellow fever they decided to close their laboratories, and rather than burn them to the ground they gave them away to the local authority. And the local authority as often as not was the British Colonial Office, who suddenly found themselves with this gift, a slightly unwanted gift. They had the insect people in the lab but the virus people were American, so it turned out that the Colonial Office in the late 1940s suddenly had to generate a lot of virus people and so they paid me to go and learn about viruses in Australia.

'And what came from viruses?'

After working for a couple of years in Uganda for the Colonial Office, my wife and I went back to Australia. This was now the mid-1950s, a time when cell culture was being used for studying animal viruses. Nobody in Australia knew very much about it, and so I was sent to the Californian Institute of Technology to learn about cell culture techniques. I had a lucky accident when I got to Cal. Tech. Often when you go on a fellowship the fellowship money doesn't come through for quite a long time, and as it happened I immediately came down with Asian 'flu and retreated to bed. They had put me in the Faculty Club and the Faculty Club bill for people who stay a short time was absolutely enormous, so after two days in bed I discovered that I'd used up two-thirds of my capital, without even feeding myself. I went in desperation to the laboratory, looking for graduate students who could tell me where I could stay for almost nothing, and one of the graduate students said, 'Well, you can stay in the house that I'm living in which is run by Matthew Meselson.' So I lived in this house at the time that he and Frank Stahl were doing a key experiment on DNA replication—which is as nice a way of being introduced to molecular biology as any.

'Meselson is a very distinguished molecular biologist.'

Yes.

'So, in effect, you fell into bed with molecular biology.'

Yes that's right. All these hot shots were a lot younger than I was and they thought of me as a doddery old man who did the washing-up.

'But you actually like doing experiments don't you?'

Oh I love it.

'What do you like about it?'

I don't quite know. An experimentalist I enormously admire, Alfred Hershey, said that what he liked about being in a lab was finding an experiment that works, and then doing it again and again and again. I rather like that. It's a curious life in the laboratory because if you do an experiment, in principle you do it because you don't know the answer. Because you don't know the answer you can't properly design the experiment. Eventually by messing around you arrive at a conclusion of how things are. You then design a set of experiments which are the most economical way of demonstrating that what you have decided is the truth, and those are the experiments you publish.

'So, do you actually like writing up the papers?'

The bit I really like is writing the introduction because I like to try and think what are the two sentences which plunge the reader *in medias res*. The bit I like least is right at the end, the discussion. The convention is that you write what you're going to do, describe how you did it, describe what the actual results are, and then say it again in the discussion, and I don't like that. I like to try and say something different, since I don't like writing something twice. I find the discussion very, very difficult and console myself with the fact that obviously people writing symphonies find the last movements terribly difficult. It's quite plain that Beethoven and Brahms found last movements very awkward.

'You enjoy doing experiments, and you describe some of your experiments as fun and clever. What do you actually mean? What are fun experiments?'

For me the fun of an experiment is like the fun of doing a lot of washing-up. You get everything ready and you make damn sure that you can't go wrong. I like to think it's something of an art if one pair of hands gets through a huge amount of work in a short period of time simply by organizing it.

'That's fun?'

I find that fun. It is fun when the experiment is over and all your samples are in a Geiger counter, and you can go away for a long weekend knowing that this faithful machine is working night and day for you for three days, and you're going to come back and look at the tape with the results that comes out of the machine. Of course, what actually happens is that you come back after this long weekend on Tuesday and find that you've not put it on automatic and so it counted one sample and then stopped. That's the down side of it. But that's part of the fun. Part of the fun is the struggle. You try to work out what it is that you think, and eventually it becomes clear and you think, 'clever old me', and you are feeling awfully pleased with yourself. That pleasure lasts for about six hours and then you're back to the normal gloom because you think, 'well, any fool could have told you that'.

'Do you think there's a lot of gloom in doing science?'

Oh yes. I think science is very difficult.

'A peculiarity of your research career is that it seems that you've worked in a large number of different areas. It's almost as if you were a prospector looking for the right places to go and mine, and you don't seem to stop for very long. Your mind seems to be quite migrant. Is that fair?'

Well, I think my body has been migrant since I worked in England and then Australia and then Uganda and then Australia and then America and then Australia and then America and then London and then America and now I'm back in England, so I really have bodily migrated a great deal.

'Why have you migrated so much?'

I've always taken the first job that is offered.

'Is that a principle?'

No, not a principle. I think he who has a wife and children has given hostages to fortune, as somebody said. So I've been trying to pay the ransom.

'OK, but your migrant mind?'

I have never been a company man who wants to develop an enormous company doing something. I've never liked teams of postdocs, and huge grant applications and stuff like that. And so whenever I worked in a subject and some field has opened up either because of, or despite of what I've done, then I feel I want to get out because it's obvious that the big empire builders take over at that point. There's no point in competing with an empire; it's very difficult for an individual to compete in science with a big empire.

'So, you've never really been in very competitive fields?'

No, it may be that when I see competition I run for my life, I'm not sure. I think the nicest way, at least for me, to do science is to go on sabbatical leave to someone else's lab, where all the ingredients are present, all the know-how is present. I work by myself on some experiment, and if I want to know how to do some little bit of it, I go and ask someone and they say, 'you do this and/or, that', and 'I can give you such and such a thing to make it work'. I work by myself but with a hidden team of helpers. That's what I really like. I don't know whether scientific empires operate that way. People don't jointly do experiments very often, not in biology. They may help each other, literally a helping hand, or provide raw materials, but the act of doing a particular experiment is nearly always done by one person, or by one person and a technician. I find the boss–technician relationship very difficult because much of the time I don't do anything because I can't think of anything to do, and to have a technician hanging round your neck having to be employed I find a great responsibility.

'You've spoken elsewhere about the fact that although you've done some very nice experiments, if you hadn't done them you are pretty sure somebody else would have.'

Oh yes. I've done very few experiments which wouldn't have been done by someone else. There's been a lot of correspondence about what would have happened if Watson and Crick hadn't divined the structure of DNA. It is clear it would have come out in a couple of years. It mightn't have had quite the impact that it did. It might have been a bit slower because they saw what it meant instantly, whereas if it had been in the hands of the crystallographers they might not have, and no-one might have read about it for a couple of years. But it's just a delay of two or three years we're talking about.

'So, you must see therefore your own work in a way as irrelevant?'

Yes, and I have the luxury of seeing most other people's work as irrelevant as well.

'Do you think that's part of the nature of science?'

I don't think it's something you can level particularly at science because I think there's a certain redundancy in all human activities.

'But I think that is different in the arts. I mean if Shakespeare hadn't written *Hamlet* . . .'

It wouldn't have been written.

'Yes, it's quite different from science. Now, if we look at some of your experiments, I think one of the characteristics of them is that, as you've already said, they open up a new field. They're also experiments which very often go against the standard view, or show that the standard view is wrong. You've worked with bacteria, and suddenly you get involved in this big evolutionary argument. Is that a conscious approach of yours?'

I don't consciously think, 'Tuesday morning, dear diary, who shall I gun down today?' I certainly don't think like that. I find myself reading some branch of the scientific literature and getting dissatisfied with it. For example, perhaps I should explain the origin of these evolutionary experiments. Having worked in molecular biology for about ten years, I suddenly found I had to make enough money for my two sons to go to medical school, and this resulted in my migration into cancer research. Cancer is, of course, a disease you have to think of in terms of the emergence of a villainous clone of cells who have a survival advantage over their sweet innocent neighbours. That set me thinking about the control of natural selection and the competition between cells, and from there it was a short step to wondering what are the forces that produce variation in cells, because it is these variants that give rise to cancer. Of course, it is a variation that has made evolution possible. The dogma has been that variation is due to random errors, if you like, typographical errors made by the DNA typist. The system that replicates DNA, has an intrinsic error rate and these errors usually are bad, but are very occasionally useful, and it is the occasional useful one that has produced the fancier forms of life like human beings, as well as the less fancy ones, over billions of years. But there have been extraordinarily few experiments to test whether variants do arise as a result of blind typographical errors.

'Chance?'

Or not—that is the question. These experiments are difficult to do because you can only investigate a phenomenon if it occurs under your very eyes. Mutations are very rare, any particular mutation you may look for may have a frequency of one in a million, or one in a thousand million. So you must look at a million or a thousand million organisms to be certain that the guys that you're looking at are the ones who are undergoing the mutation. Therefore all these experiments have to be done with creatures like bacteria and yeasts. They cannot be done with mice. There's been one experiment done with a million mice to test radiation sensitivity, it's called the Megamouse Experiment, and I don't think anyone will do another experiment like that. But you can easily work with a thousand million bacteria, it costs you ten pence or something like that. A bunch of experiments were done in the late 1940s and the 1950s to ask whether mutations are continually arising in bacteria, and the answer was yes. My lab went back to looking at these experiments, and so did a few other people, and we found that if you subject bacteria to a selection procedure which isn't immediately lethal to anyone who doesn't pass the test, then in fact a few do work out how to survive. So you then say, would they have produced those mutations if you'd set them another test. Are they only producing answers to the test they're being set, or are they just producing answers to all possible tests? And it turns out, under certain conditions, that they're only answering the test that you set.

'That was a big trauma for those working in evolution.'

Yes. It's not anti-Darwin, who was Lamarckian actually, but it is anti- the twentieth-century neo-Darwinists, and so, naturally, the neo-Darwinists have been very upset about our experiments. I think maybe they're particularly upset because they're nearly all theoretical biologists and have lived happily working out why bluetits have twelve eggs instead of ten eggs in their nest, or what happens if you turn the nest upside down and that kind of thing. They always work out some reason why the little bird does what it should be doing. This is extremely interesting and sometimes very, very important, but they're not really experiments because whatever the result, you work out the reason for it. They're not attuned to the idea of doing an experiment to see whether a creature can do something which you think it shouldn't do. They try to work out reasons for the experiment to be wrong, which is right and proper since you give up cherished beliefs only with reluctance.

'What did you actually feel when the experiment worked?'

I don't think there was any particular moment when I leapt unclothed from the bathtub or anything. I would worry about it a great deal; think, 'what experiment can I do which asks more incisively the answers to the kinds of questions I'm thinking about?'. I did a lot of jogging at the time, around Boston, and when I was getting fitter and could jog without thinking about running and think instead about experiments, I would say to myself, 'Will I, in the next ten miles, clear up my mind on this?', and then at the end of ten miles I was feeling physically very well but I hadn't cleared up my mind at all. We wrote a paper about this which was published in 1988, and since then a whole lot of other experiments have been done. It is now pretty solid that something is going on in bacteria so that they produce useful mutations and don't seem to produce useless ones. Now, it may be that they're producing all kinds and have a way of getting rid of the ones that are useless but we don't know the mechanism.

'Do you think this is the right way for science to go forward? For someone like you to come along and upset the applecart? Was there tremendous resistance to your ideas?'

Oh yes, and still is. I don't know what is the proper density of upsetters of applecarts. You want to have some, but you cannot have everybody doing it.

'But this goes against the image of scientists being open-minded, willing to change their minds, excited to hear new results.'

I don't know that scientists are any more willing than anyone else to change their minds. If you've worked on something and feel that you see it clearly, you don't like to be told that it is not right. Einstein didn't like quantum electrodynamics and, to his dying day, felt that it was intolerable and objectionable because it was so unbeautiful. I think there are all kinds of reasons for being conservative which are perfectly honourable ones. And also, you don't want everyone rushing off in all different directions at once because you then have chaos, you want a certain dampening of the oscillations of excitability.

'You said you got into cancer research really because you needed the money. I mean that's a bit odd. Then how does one begin to think about cancer research, as it were, coming into it from the outside?'

Leaving aside the money?

'Leaving aside the money.'

Well, I think several times I've been very lucky in my life because these transitions have occurred when I was old enough not to feel threatened, or when I was working in a place which didn't require that I write grant applications, where I could do whatever the hell I liked. When I moved into cancer research I decided that what I should do is read. I love libraries, the slight buzz that the fluorescent lights give off, and the smell of gently rotting books in the stacks and the complete isolation from the world, no sound of traffic. You sit there getting more and more desiccated, a shrivelled up old person. Reading books is very satisfactory. So, I read. I suppose, this was about 1971, '72, and I think at that point you could read all of cancer research in a year. It would be a few thousand papers, and I think I read them all. As you know, some papers take a lot of reading, others take almost none at all. It was, in a way, somewhat unintellectual stuff because I don't think at any point did I feel that when reading I would suddenly clutch my brow and give a shout of joy as I realized some deep truth that escaped everyone else. I don't think in science that happens. You've got to grind away a lot more than that, and anyway I didn't feel in cancer that there was a deep truth waiting to be discovered, I just wanted to get a clear overview of the state of play in all its branches.

'Thinking back, what are you best pleased with of your work?'

I don't think there's any particular thing that I am pleased with. I'm pleased when I look back at the various things I've done which are distinct and identifiable. I make that qualification because I think that a lot of the things that scientists do are mush and noise, general noise, and they wouldn't count them when approaching the pearly gates. They wouldn't hand these in as possible basis for a credit card. I can think of things I have done that please me, not so much because of the things themselves as the fact that I got it right. I know one thing I got wrong, which was not very important, in 1951, but it's ceased to be interesting. One of the things that happens to burning questions in science is that with the passage of time they just vanish, their interestingness vanishes. So as you go through the literature you see these trunkless legs of stone and think, ah! this will happen to yourself and your own work. Science goes on and nobody cares.

RICHARD LEWONTIN
was born in 1929 and is the Alexander Agassiz Professor
of Zoology and Biology at Harvard University.

CHAPTER 11

Not all in the genes

Richard Lewontin
Evolutionary biologist

MOST scientists, myself included, are committed reductionists. Modern physics seems to demonstrate beyond reasonable doubt that all the complexities of the cosmos will ultimately be explained in terms of a few forces and particles. Similarly, modern biology attempts to reduce its mysteries to explanations at the molecular level. Richard Lewontin is one of the few who is not in sympathy with such endeavours.

Despite being a distinguished geneticist and, until he resigned, a member of the US National Academy of Sciences, he is a vigorous opponent of reductionism. He is particularly hostile to the extention of the reductionist approach to justify biological determinism—the idea that much of the behaviour of animals, including man, is controlled by genes. Suggestions of the kind made during the so-called IQ Debate, that attributes such as intelligence are largely inherited and racially variable, appal him. As a vociferous Marxist and critic of much that the scientific establishment holds dear, one might expect him to be uncomfortable in a professorship at one of the United States' most august universities. This was one of the reasons I wanted to talk to him, but I also wanted to understand how an anti-reductionist thinks.

I don't think the world is made up of natural little bits and pieces that fit together in some natural way and bring to whole objects their own properties. I think the properties of the bits and pieces we divide the world up into are properties they acquire in actually being parts of wholes. When scientists break up the world into bits and pieces, they think they're breaking things up into some natural entities, but they're not. They're breaking them up into the bits and pieces that they see in the world by their own training; by their own ideology; by the whole way in which they've been taught to see the world. But that doesn't mean those bits and pieces are the real bits and pieces. So, an anti-reductionist is someone who believes that in order to understand processes of nature, you must not import into them some artificial notion of 'real' bits and pieces. Now, one must also remember that things that are generally true are not at the same time universally true. There are bits and pieces which in every sense are functionally independent of other bits and pieces, but an anti-reductionist's stand has to be one of being suspicious.

'Do you apply this across the whole of science? I mean, would you include chemistry, biology and physics? What about atoms for example ?'

I certainly would. The claim of physicists that they've almost got down to the ultimate bits and pieces has been going on since the Greeks. I mean, atoms—those things which can't be cut into pieces—were then cut into pieces. So then there were the elementary particles. But then the elementary particles turned out to be not quite so elementary and now, when I ring up my physicist friends and say, 'Well, have you finally got the bits and pieces, the quarks?' they say, 'Oh sure, this time we know for sure.' But you know and I know that they haven't. And I sometimes think that high-energy accelerators which break things up into bits and pieces are like sledge hammers that break up bricks. That is to say, every time you hit the brick you break it up into bits and pieces, but that doesn't mean that those are real bits and pieces. So I'm sceptical even at the level of physics.

'But I want to stick to molecular biology, or biology, in general, where I think most of us would say that over the last twenty years there's been an amazing revolution which has transformed biology—and that's the reductionist approach of explaining biological processes in terms of activity at the molecular level. Would you agree with that?'

Well, I think it's revolutionized a lot of biological practice, and it's made possible explanations of a lot of things. It hasn't revolutionized the general viewpoint of biology, which has been completely a mechanical reductionist viewpoint since the nineteenth century. But the thing I want to say about molecular biology is that science is supposed to be the art of the soluble. What that means is that smart people who become scientists are not crazy enough to spend their lives trying to solve really hard problems. You're much better off doing molecular biology than you are trying to find a way to cope with schistosomiasis, a disease which affects millions of people in the Third World. If you're a molecular biologist, you and your friends can decide (a) what the problem is, and (b) when it's solved. But in the case of schistosomiasis you can't *decide* what the problem is, because it's given to you; and you can't decide that you've solved it if people still have the disease. In general, what biologists do is to act like like dentists: to follow when they're drilling the soft and decayed parts, and leave the hard parts behind.

Molecular biologists have had great success in telling us what the physical chemical nature of the gene is, and what the structure of the genome is. They've had great success in explaining how some genes are turned on and turned off, and things of that kind. But they haven't done very well so far, either in the nasty problem of development, or the nasty problem of the central nervous system. And I think the problem of development, and the problem of the central nervous system have one important thing in common, and that is you're dealing with immensely complex physical and dynamic interactions among a lot of bits and pieces, and molecular biology simply doesn't have the kind of structure of explanation which will put everything in context. It'll talk about little bits and pieces by themselves, but the problem is to put them together, and what worries me about molecular biology is this attitude, 'it's all in the genes', and it's not.

'One of the features of your, I don't want to call it hostility, but your suspicion of the reductionist approach, I think is linked to your real hostility, and that's to the idea of biological determinism. Would that be fair?'

I think that's not quite right, Lewis, although they're connected. I have always done things that I would call non-reductionist, and even dialectical, for my experimental work from when I began as a graduate student at the age of twenty-two. I didn't think about biological determinism then. In fact, that was just after the Second World War when biological determinism was at an all-time low as a consequence of the Nazis and so on; people were very much environmentalist and anti-racist and so on. But, nevertheless, I had an intellectual commitment—I don't know where it came from, and it's not interesting to know where it came from—an intellectual commitment, from a personal standpoint, to seeing things in context. Now, that made me more sensitive to biological determinism. I'm also sensitive to it for purely political reasons, and the consequence is that as I began to struggle against biological determinism I began to re-examine a lot of biology that I had previously taken for granted.

'But in the more, as it were, popular concept of biological determinism, there is the question of how much of our own behaviour and that of other animals is determined by their genes, and you've certainly been involved in the IQ and inheritance debate.'

As one example.

'As one example. Could you explain where you stand on that?'

Well, the problem of IQ in particular is a complex one because it has so many errors piled one on top of the other. There are purely methodological errors, and I don't want to talk about those, but there are also conceptual problems. That is, is IQ a thing which can be caused by genes, or is it a social construct? If it's a social construct, then we have an interesting problem, the heritability of a social construct.

'Can I interrupt and ask the question in a slightly more simple-minded way—and I realize there are technical problems with this—but do you think there are features related to human intelligence which are inherited?'

Of course there are. You and I are having a chat here, and no two monkeys will ever do that, and the reason is that our genes differ from monkeys. All the claims that people are talking to monkeys by computers may or may not be true, but let's not forget that it was we who made the computers and not the monkeys. So, yes, in that sense.

'And individual differences?'

The question of individual differences in intelligence is partly a conceptual one. What do we mean by individual differences in intelligence? Do we mean that I can do some things, like turning screwdrivers and disconnecting telephones that you can't do. That's certainly true. Does it mean that despite my deep desire, I cannot play the violin like Yehudi Menuhin, and probably couldn't have played the violin like Yehudi Menuhin even if I'd started at the age of three? I'm sure—I can't prove this, incidentally, this is very unscientific of me—I'm sure that Yehudi Menuhin has nerve connections and muscular relations that I don't have, and probably would not have had, in the absence of great knowledge about how to make development in a human being. That is not the same thing as saying, however, that Yehudi Menuhin and I are different in this respect because our genes are different. We must remember that a good deal of differences in development between individuals do not lie either at the level of gene differences, or at the level of what we conventionally call environmental experience, but lie in a realm which is still under investigation, and which Waddington called developmental 'noise'. This essentially means growth of neuronal pathways which is not a consequence of different temperature, nor a consequence of different genes, but of some kinds of intimate cellular events that we now call random, and for all I know, Yehudi Menuhin and I differ by the neural differences that arose by these random nerve differences.

'But equally they could have arisen by genetic differences?'

They could have been given by God, Lewis. The issue is not what could be, but what is.

'And it's also what's plausible, because we know that genes can alter neural connections, so isn't it plausible?'

Of course it's plausible. But the struggle against biological determinism is not a struggle against an implausible claim, it's a question of what *is*, not what *might* be. And this is very important because, knowing the ways genes operate, there is almost nothing that is impossible in any genetic story that you tell—the problem is what's *true*?

'Is your attitude towards social determinism somehow linked to a particular political viewpoint?'

Of course it is, but it's not linked in the vulgar way that people like to say. What politics does, or ought to do, is to sensitize you, first of all, to certain world views, and, second, to make you suspicious. And these suspicions, how are they related to politics? I would say if I had to distinguish between the ideology of the left and the ideology of the right on questions of biology, it's that the right has a commitment to the fixity of things, and the left has a commitment to the changeability of things, and the onus is really not symmetrical. The people on the left spend their lives trying to struggle to change the world . . . but suppose it were true that differences in status, wealth and power in the world were unchangeable and entirely the products of genes, then people on the left would be out of their minds to spend all that energy, time, money, and commitment, psychic commitment, to try to change something unchangeable. What's the sense in struggling against a fact of nature? So, it's very important for those of us who say that the world can be changed and want it to be changed, to know whether it can be changed. That's not true for the political right. Suppose it were true that we're right, and that all those things can be changed, and are not ineluctably in our genes. It's still a perfectly coherent political programme for the right to say, 'but we won't change them'. So I would say it's the right who has no particular interest in knowing the truth of the matter, because it can carry on its programme irrespective of the truth, whereas the left has a deep commitment to knowing what's true.

Let's take the case of IQ in particular, because I think it's a very important point. Please note that studies of the hereditability of IQ are not studies on whether IQ is changeable, and that is one of the neatest bits of intellectual sleight of hand that biological determinists have pulled. They claim to want to know, for example, (I'm referring here to Jensen's famous article) 'How much can we boost IQ in scholastic achievement?' Now, as an ordinary person, you would think, 'well, if my question is, "How much can I change something?", then I ought to do an experiment to see how much I can change it.' But no, instead of doing that, what they have done is to say, 'Let's see whether there are genes involved.' Now, why have they done that? Because they make the error of thinking that if they can show that genes are involved, that's equivalent to saying, things are not changeable. But as all geneticists know, one has nothing to do with the other. Genes can indeed be involved in any way you like. Nevertheless, by a change in the environment one changes things.

'But are you implying somewhere, and I think you have in your writing, that quite a lot of science is actually—the nature of science itself—is determined by these political prejudices?'

Yes, indeed. Scientific ideas certainly come out of views of society, but not always in some vulgar, political way. Scientists see nature reflected in the mirror of social relations. People are first of all social beings and only secondly scientists. We talk about social Darwinism as if social theory got its ideas of survival of the fittest from Darwin but, of course, the reverse is true. Darwin got his ideas about selection from the social milieu in which he lived. Darwin read the Scottish economists. He spent an hour every day reading the shares lists in the newspaper because his income came from investments in shares. Darwin understood very well notions of selection, survival of the fittest, differential reproduction of capital, and so on. That's where he got his ideas from. His grandfather was a self-made man. He was part of that circle of Midland industrialists, including Wedgewood and other self-made men. Darwin lived in a time of dramatic and radical social change, and at the end of the end of the eighteenth and beginning of the nineteenth century a very important change took place in general view. Before that time, change was regarded as exceptional, and constancy was regarded as the normal circumstance. So we had theories of the extinction of organisms because volcanoes erupted, or floods came. The present view that the natural aspect of the world is that it's constantly changing—which is what we mean by evolutionism, a notion incorporated in evolutionary cosmology by Kant and Laplace in the end of the eighteenth century—that's a bourgeois notion. Now, the commitment to an evolutionary world view is a commitment to a social view, and it's only at the very end of the process that it comes into biology. So, there's an example where social views are very important.

'You're almost a relativist.'

I'm not quite sure I know what a relativist is.

'Well, that is saying that science isn't very special knowledge, but it's just merely another manifestation of the social environment in which we live.'

But that's because you're a reductionist, and therefore you want to say, either it's objective knowledge, or it's a manifestation of social milieu. You have to be more dialectical in your thinking, Lewis.

'What do you mean by dialectical?'

Well, let me try to exemplify it, if I may. It is certainly the case that irrespective of social views about the world, the distance to the moon is the distance to the moon, and nothing about ideology can change that. There are facts of life, and in that sense I'm a realist. What is important is to know how people pose questions about the world, and the dialectical point is that the answers they find out are contained in large part in the questions that they ask, because they transform the problem by the very question they ask. Let me give you an example of a terribly undialectical viewpoint which is common in the study of evolution, and which we oppose. People think they can study the evolution of morphology. A famous question which can't be answered, but it's a question, is why did certain dinosaurs have big bony plates along their back? Four answers have been offered, and none of them may be right. One is that they were a sexual recognition signal so that the males could recognize the females of their own species. Another was that they interfered with predators biting them, because they got caught in the predators' teeth, so to speak. A third was that they made the animals look bigger, and so they scared off predators. A fourth is that they were thermoregulatory and cooled the animals.

Now, please notice that all four of those answers depend critically on the behaviour of the animals. You cannot know anything about whether it's good to have a thermoregulator on your back unless you know whether you go out in the sun or not. You can't know whether sexual recognition is important or not important unless you know what the courting patterns were. There are no morphological problems of evolution that can be detached from the actual life activities of the organism. So, it's a dialectical principle of evolution that you can't just break up the organism into arbitrary bits and pieces. You have to see the whole thing and let it somehow, as the process of investigation goes on, tell you how to break it up. Let's say, we don't know how to make the rules of dynamical change in evolution without knowing the objects that those rules describe, but we can't know how to describe the objects without knowing what the rules are. So, it's a dialectical principle, if you like, that both the rules and the objects about which the rules are made have to be simultaneously tried, and fit.

'There's a curious paradox I see in your position. Your thoughts are radical, in the sense that they don't fit into the conventional view of many scientists, yet here you are, a professor at Harvard, an anti-establishment figure, in a way, in a very establishment university.'

I have to make a living like everyone else, that's point one. Point two, in order to have any success at all in raising people's consciousness about thinking in ways that they don't ordinarily think, one has to have legitimacy. Legitimacy is the most important problem, and a metaphor that occurs to me, a very bourgeois metaphor, is the bank account. When one tries to convince people to think in ways other than they usually do, one is drawing money out of a bank, and you have to put money back in the bank because you soon have an overdraft, and these particular bankers don't give you a very big overdraft, so being a professor at Harvard is very useful. It's very important for me to be able to speak from a platform which implies legitimacy, so that then people have to say, 'Well, he can't be entirely a crackpot, after all he's a professor at Harvard, so maybe we ought to pay some attention to what he says.'

'Do you feel a bit alienated, then, from the community?'

Of course I feel alienated from it. I have little or nothing in common with any of my colleagues. We take the opposite point of view on almost everything. I try to get the department to listen to, and to even give a vote to, the non-tenured members of the faculty. My colleagues say, no, they, as older and wiser people, really cannot share power. On every issue we struggle. We struggle about the way to organize the work in the laboratory. We struggle about whether I should be the boss or not. Who is responsible, who is to make decisions? Of course, most of my colleagues live in a social circle to which I don't belong. I don't have a single personal friend on the faculty of Harvard University, with the exception of Steve Gould and Dick Levins, both of whom are Marxists.

'You actually resigned from the National Academy of Sciences, if I remember.'
Yes, I did.

'Can I ask you on what grounds? You may not want to talk about it.'
Oh no, I'm delighted to talk about it. The National Academy of Sciences

represents politically an institution that I oppose and that I try to de-legitimize, and that is the institution of giving honours and creating élites. I am against honorary degrees, I'm against medals of science, I'm against prizes. I will certainly never win a Nobel prize, so it's easy for me to say that were I to win a Nobel prize I wouldn't accept it. I certainly have turned down honorary degrees.

'Why, why?'

Because I believe that the work that some of us are trying to do, which is to try to understand something about the world, is distorted and warped by the existence of prizes. I said before in this interview, perhaps it won't be on the tape, so I'll say it again, that very able people in science, call it smart if you like, are not crazy enough to devote their lives to really hard problems because fame resides in solving problems. So one of the things that these élite institutions do is to make sure that people spend their lives—motivated by the desire for fame and for élite status—working on problems that they can solve and get famous by. Our graduate students are constantly being bombarded by the idea that maybe some day they'll win a prize, some day they'll become a fellow of the Royal Society. I see people whose lives are blighted by failing to be elected fellows of the Royal. These people have no right to make people's lives miserable in that way.

'You are an anti-establishment figure, that's absolutely clear. Do you enjoy that position?'

Well, Lewis, I sometimes enjoy a fight, and sometimes I don't. I am not, I have to say, one of those people whose temperament makes him struggle and fight at every opportunity, I really am not. Many struggles and fights I find deeply temperamentally disturbing. I do them because I must.

'Where do you get your pleasure in science from then? Do you get pleasure in science?'

I get pleasure and I get pain, as everyone does in science. I get pleasure in science from solving problems. I do both theoretical and . . . or I should say, I do theoretical work and I administer experimental work. Like all fairly successful scientists, I soon stopped actually doing experimental work. I get a lot of pleasure when someone in the lab discovers something that I think is true about the world, and I get a lot of pleasure when I do a theoretical problem and I come out with a theoretical solution to that problem. So my pleasure in science, I would say, is exactly what every scientist says, my pleasure in science is finding out something that's true. My claim is that most of them are not telling the truth. That most of their pleasure in science comes out not from knowing what's true, but from claiming something that other people think is grand.

ANTONIO GARCIA BELLIDO
*was born in 1936 and is Professor at Centro de Biologia
Molecular, University of Madrid.*

CHAPTER 12

In agreement with nature

∽⸺⸺⸻

Antonio Garcia Bellido
Developmental biologist

SPAIN does not come high up on the list of countries which have made a major contribution to the development of modern biology. With the outstanding exception of the nineteenth-century anatomist, Cajal, who won a Nobel Prize for mapping neural connections in the brain, Spain has no strong tradition in biological research.

It was in this scientific vacuum that Antonio Garcia Bellido set about first becoming a developmental biologist, and then founding, from scratch, what is now a flourishing and influential Spanish school of research. To do so, he had to study in Britain, where he did insect physiology with Sir Vincent Wigglesworth at Cambridge, then in Switzerland, where he trained under the geneticist Ernst Hadorn in Zurich, and in the United States, where he worked with E. B. Lewis and A. H. Sturtevant at the California Institute of Technology.

Why had science been so neglected in his own country?

———

If we go back a long way, from the eighth century to the sixteenth century, I would say Spain was very good at science. However, we missed the introduction of scientific experimentation, around the time of the Renaissance, which was introduced by Galileo into central Europe.

'Why did Spain miss out there?'

Fundamentally because after Philip the Second's II's law of 1580 there were fears of the invasion of Protestant ideas into Spain and Spain just closesd itself in. The Pyrenees became a barrier. There was very strong control over what was being taught in the universities. Whereas, before, Spanish scientists went to teach in many other countries, from that moment in 1580, Spain became isolated, so to speak, from the rest of the world, and from then on the state took over.

'How has that affected modern science in Spain?'

Very much, Because science, in order to be alive, has to be maintained, well fed. In the moment something decays it is very difficult to recover it. So it's not a short-term effort that one needs but a long-term effort. Spain is making a lot of effort just to recover the idea that science is interesting, is worthwhile, that it provides power in the long run. Members of our society have to feel in their bones that this is true. We had lost this commitment, and it is going to take a long time for us to recover it.

'But you have had famous nineteenth-century scientists like Ramon Cajal. Is he an exception?'

Right. Cajal is, yes, he's outstanding. Most of the Spanish creation, most of the Spanish originality, the thing that you admire about Spain, is the personalities, for they are exceptions to the rules. They are individuals that are capable of breaking with the central dogmatism, religious or philosophical, and yet be capable of creating something. It's not just simply that they are unorthodox. Cajal is the paradigmatic personality in this class. Cajal is unique and breaks out of very boring, poor scientific surroundings and builds by clear ideas. He had a very strong personality, and was extremely hard-working.

'Did Cajal establish a neuroanatomical tradition in Spain?'

In Cajal's time it became obvious to young medical students that what they had in Cajal was a model person. Therefore the best students tried to work with him. So he was building a classical school around him that was suddenly discontinued with the Spanish Civil War. Many of the key scientists that had personal contact with Cajal left. The people that remained lacked the originality and 'avant-gardism' of Cajal's time.

'Would you say the recovery of science in Spain really only began at the end of Franco's rule?'

No, because most of the current science in Spain had started already in Franco's time. It did not start because of Franco, it started because there were scholars who felt they had to do science. People started visiting other countries. There was a feeling that Spain had to recover. This happened at the level of individuals, not at the level of the state or politicians. When I went to England there was no such thing as a fellowship to go to England, and when I went to Zurich in Switzerland I had to find a private foundation to pay me for going there because there was no government support. So, science started in Spain despite Franco.

'How is science perceived now within Spain? Is it a prestigious subject?'

I think it's an unknown subject. Obviously the academic community is closer to the politicians. For example, the Minister of Research is a scientist himself. But I don't think that this goes over to the public. The public is rather ignorant, maybe because there haven't been reports of science in the Spanish newspapers. The public feels that science is something which is being done abroad. Science is not part of the conversation of the people in the street. But now a lot of money is put into science. There is a real revolution. Our professors in the university (they don't do research in other countries) do research in Spain. Out of a desire to emulate, and even envy, they want to do as much research as professional researchers. And there is money now. Things are changing. But obviously this taking off takes so long. It took Germany at least thirty years to start doing good science after the Second World War and now it's doing it. Let's hope that Spain will do it maybe sometime soon.

'What about the relations between the Church and science in modern Spain?'

None, I mean simply none. It doesn't fear science and doesn't help science.

'It ignores it?'

Ignores it. So, in Spain at this moment a situation like that of the United States in relation to, for example, creationism in contrast to Darwinian evolution, could not occur, because the notion of evolution is completely embedded in the mentality of everyone. So, there is no contesting, like within the puritanical Anglo-Saxon societies. In Spain the Church does not intervene. You don't have to worry about that, you can say that, well, evolution is a scientific notion and therefore you have to explain it as you can explain now the Copernican notion.

'Did you become interested in science as a child?

'Well, you can never say that. But I would say relatively early. Both my father and mother are scholars in the humanities, in archaeology and in Greek and Latin philology.

'So, they were academics?'

Yes. The atmosphere of the family—I happen to be the first born—was academic. The only worthwhile aim to be was to be a scholar and an academic person.

'Was this in Madrid?'

Yes, Madrid. The atmosphere was permeated by the humanities. But my father, whom I do admire very much, happened to have a general interest in knowledge as a whole. He had a very good private library which included books on mathematics, physics and biology. At thirteen or fourteen I was reading these books, thinking they were needed for my education. In the same way, for example, you have to read the *Iliad*; it's part of your general cultural education. You have to do it. And in this reading I very soon became polarized towards biology. It was the kind of thing that appealed to me.

'Do you remember why?'

Reading the classics, like Darwin and German translations of books on development, I became very strongly interested in development. For me it was very clear that the real problem was development—how an organism is made. And from then on I made my programme and started fulfilling it—to do biology, namely to study natural sciences. All the family was against that notion, they wanted me to become a medical doctor, because it was the only decent profession, or to be an agricultural engineer, that was rather decent too. But to be someone just racing after butterflies—no.

'But that's a puzzle because, after all, your parents weren't working in practical areas.'

They are conservative. I remember my father saying to me, and I insist he was a very intelligent person, 'Look, you do medicine. And when you are finished with the medicine, then you can do research, since then you will have a profession, a social status.' But things were changing and youth was smelling that you could do things beyond the normal routes. I just followed my ambition. I had dared to do something that at that time was highly unorthodox.

'But was there any school of developmental biology where you could study?'

No, nothing, nothing. I had to explain that I needed to study natural history, botany, zoology and systematics, comparative anatomy, and all the classical subjects.

'Why did you choose insects?'

All right, because it was clear at the end that the only answer to development had got to be genetics.

'So, you had come to that all on your own?'

Very clear.

'But Wigglesworth wasn't a geneticist.'

No, he was not a geneticist. But in order to understand a subject you cannot approach it with one single knife, you have to use as many knives as you can. Insect physiology was fundamental and also I had to have genetics. Without them I would go nowhere. I had a programme, as I told you, I wanted to do it, one step after another. I knew with whom I had to do it, so I went to Wigglesworth in Cambridge and learned about the insects and *Drosophila*. Then I came back and I started my Ph.D. I did my Ph.D. alone, without a supervisor.

'It's remarkable that you did it all on your own without either supervision or even psychological support. What helped you through that?'

Well, I didn't have psychological support. I couldn't have a tutor because nobody knew about basic genetics and I wanted to do genetics. I could read German and I speak German well, so I was simply following the rules I found in the publications of Ernst Hadorn with whom I was in correspondence.

'I like the idea of doing a Ph.D. by correspondence course.'

No, no. It's not exactly that, but from time to time I would just write to tell him my results. There came a moment, I remember very well, when I wrote to say I would like to come to his lab as a postdoc. Then he replied that he wanted to talk with me before committing himself. So I went to Zurich, trembling in all my arms, to have a meeting with the Professor. He didn't even bother to receive me the first day, and put me with one of his students. For three days

I was explaining my Ph.D. work. In the end Hadorn came to me and said that he was very pleased to meet me. He invited me into his office and said, 'Yes, you can have position here whenever you want to come.'

'But when you were in Spain, were there not times when you felt very isolated?'

No, no. All the time the drive was on the inside. I coped with it. I enjoyed doing it. But then I learned something, that after all the scientific isolation of Spain was not so dramatic. The only thing that we had to establish was a kind of conversation; knowing how to converse on the same wavelength and with the same rules. And I felt I was already well read so I could speak with the people about theories and about science. This gave me the confidence that the isolation is not insurmountable.

'Do you think different countries have different styles of doing science?'

Wigglesworth was very much of a gentleman. At that time I had not even started my Ph.D., but he, I don't know why, treated me with enormous cordiality. He showed me around Cambridge, the old Cambridge, and Caius College and so on, and I couldn't understand why, because he had very serious things to do. So I have a tremendous respect for him.

'How was Hadorn by comparison?'

Hadorn was a very different person. So I would say Wigglesworth had an aristocratic mentality. Hadorn had a much more efficient-minded mentality, with a tremendous personality. He was a grandiose professor. He really controlled the institute by his personality and by his science.

'A more hierarchic laboratory?'

Yes. The German style of the time. And in the United States exactly the opposite—a totally easy-going style of science. But from the point of view of working in science the style is more or less universal.

'What did you come back to, in Spain, then? After all you'd done it all originally on your own.'

I am a Spaniard, I happen to like Spain. I think I owe something to Spain and I would like to contribute, if possible, to the future respect and prestige of Spain. It's a natural thing that comes from a more primitive education.

'In Spain genetics and development are really strong now, and there are some important groups. Would it be fair to say that it really all spread from your initial efforts?'

Well, it would be too blunt to say that, but nevertheless I would say the answer is, fundamentally, yes. Many of the scientists were my students, and others because they went into the field because of suggestions of mine.

'How do you think one should measure success in science?'

Oh it's very difficult to measure. The one theory that really drives the scientist is the enjoyment of discovery. I mean the pleasure of finding something that explains the phenomena, that may have a universal value. That is, it may go well beyond the understanding of the particular experiment. This feeling is a feeling similar in many respects to the feeling of a creative artist who has finished a painting or a musical composition. It's a feeling of having been in agreement with somebody, with nature in this particular case, in resonance. You have grasped something which is hidden from the rest of the world. That, I think, is personal success, the one that really drives you. The other comes, usually with age, not to the person but to the society. When they realize that all this effort, is valuable, since it is acknowledged in other countries. This must be something special, it's not a trivial thing, and that normally suggests back to society that it has been worthwhile. It's a way of being as successful, just as you would want your music presented in Moscow, or producing a car that is being sold in New Zealand.

'You would rate recognition by other scientists as being terribly important, your work being widely quoted?'

Science now progresses at such a speed that papers ten years old are already classic and very often even ignored because the people don't have to fight with established ideas any longer. They are working on precise details. So, I would say that the pleasure of recognition would be two colleagues, old colleagues talking, chatting together, I mean it's personal, private, as opposed to a social and popular, success.

'But there are scientists who, I certainly know, get very upset if you don't quote their work in a particular context.'

Sure. If they do say that, it's because they are very upset. But this is compatible with being a very good scientist. For example, the great geneticist, Müller was extremely sensitive with respect to recognition. But I don't think it is a general feeling. Good scientists, I mean the people I do happen to admire and respect, for me are natural. They have something to say, it's something which is going still to be true, or is to be remembered, ten years later. What has happened in the last year can be very short-lived.

'What do you think about some big laboratories, where the chief scientist publishes a paper almost every two weeks because his or her name is on everything that comes out from the laboratory?'

You see our science is less of a personal endeavour, of an intellectual endeavour, than it used to be. Now it's a question of facts. And to discover these facts you are in a huge laboratory—like an industry—and the facts are coming out fast. You have to publish the work fast. Also, the students and postdocs, the people that actually do the work, need these publications. So the whole endeavour adds up to an increasing amount of information. The problem is that very often the papers overlap in more than 50 per cent of their content. There is a tendency to publish the data, not in one single paper, but 20 per cent in one paper along with 20 per cent which is covered in other paper. That can only be explained, not because of scientific reasons, but because of the necessity to promote the laboratory, a lot of publications eventually feeds back into having more grants and more money. It's a kind of a wheel which we are unconsciously feeding and I think it's a mistake.

'Do you worry about the competition?'

No, not at all, nothing. I ignore the competition totally, otherwise I will be killed. I have a saying 'Let's fight because, though the battle is already lost, we may win'. In the conditions of Spain if we try to do the same as others with just the same procedure we are always going to be too late. So we have to be bold, we have to be daring. We have to try to propose things which are not the obvious things and then change things. We must be honest of course. But unless we do this we have lost the battle. I do believe in the individual in science. Maybe it's a romantic notion. But I do believe in it, as distinct to the social aspect of science which is that you accumulate knowledge and the truth may be somewhere between the opinion of x and y and z. I think that it's important that somebody with a name contribute, and I would like to do so.

'Do you see yourself as a typical Spaniard, if there is such a thing?'

Yes, as you say, if there is such a thing. I don't think there is such a thing as a typical Spaniard. But perhaps in some ways it's like in taxonomy where we have a species, but in this case it would be the individual, and it's true that there are general characteristics running in families. There must be something very Spanish. I would rather say that I have a Mediterranean temperament and this is physiology. It's physiology, the hormones.

'Do you really think it's physiology?'

It is the physiology. It's the number of hours of daylight in the year, and the climate. But education has made me more than a Spaniard. To be more precise, I have to go to the level of genera. I have been made Castillian. I am Castillian as opposed to Catalan. The Castillian has a tradition of liking ideas, poetry, mysticism and is less pragmatic than a Catalan. Maybe the difference, in my particular case, would be intellectual education. I am not so much a Spaniard when it comes to the mental discipline that I have learned in countries like England and Switzerland. I also have a very important German component in my education. A major feeling is that rigour is crucial and that there are correct procedures for doing things. That there is a logic behind things which can be fished out. On the other hand, intuition and feeling are important. I am not like a general Spaniard, who would be more of the intuitive and dogmatic, especially the Castillian with a dogmatic mentality, less rational than a German would be. But I think the combination is an ideal one and I may even have it [laughs].

EUREKA

SIR JAMES BLACK
was born in 1924 and is Emeritus Professor of
Analytical Pharmacology at King's College School of
Medicine, London.

CHAPTER 13

Daydreaming molecules

❦

Sir James Black
Pharmacologist

BETA-blockers have revolutionized the treatment of coronary heart disease and high blood pressure. Cimetidine, better known by the brand name, Tagamet®, has saved thousands of patients with stomach ulcers from surgery. To have discovered just one of these drugs would have been a lifetime's achievement. Sir James Black invented them both and was awarded the Nobel prize in medicine in 1988. How did he do it?

As recently as the 1960s, looking for new drugs was largely a matter of trawling randomly for compounds that might do something useful. Black took a much more direct approach. He set out to find compounds that would block the action of particular chemical signals. Patients with angina get a searing pain when their heart starts to beat faster. This is because the damaged coronary arteries cannot supply the heart muscle with the extra oxygen it needs. Because an increase in heart rate is a response to adrenaline, the chemical released by the nervous system in response to stress, Black reasoned that the symptoms could be relieved by blocking this reaction. This is precisely what he succeeded in doing. The beta-blocker, propranolol, works by binding to the sites in the heart, the so-called beta-receptors, which would otherwise be stimulated by adrenaline. Cimetidine, the anti-ulcer drug, works on a similar principle.

Sir James's original training was in medicine, and he switched to research and physiology after qualifying. Since then he has moved backwards and forwards between universities and the pharmaceutical industry. He now runs a research foundation which is funded by the industry, but guarantees him almost complete independence. Had he embarked on such a career with a burning desire to help humanity?

———

I am certain that it was entirely intellectual.

'That's strange because you've done an enormous amount that could be regarded as compassionate. Your drugs have saved probably more lives, and people have been quoted as saying this, than any doctor could possibly hope to have. And yet that wasn't your aim?'

No. One strong feeling I have at this stage of my life is that there are certain things you may not seek directly. There's a kind of obliqueness necessary in life. An obvious one would be happiness. I think that in terms of trying to help people, I'm not sure that I would have got anywhere if that had been my drive.

'So, what was your drive, then, when you studied medicine?'

What was plain at the end of my medical course was that I was in rebellion against the way medicine was practised. I found some of the ways that patients were being treated distasteful. It seemed to me that a kind of pseudo-science was being practised. Science, as I understand it, is about laws, about generalities, but the practice of medicine has to be anti-science in the sense that it has to be passionately concerned for the individual. This was how I felt. I was very young at the time, I admit. This was something with which I was unhappy, and so I decided that physiology was something which left a nice taste in the mouth, like listening to Mozart, and so I went into physiology.

'But that's not pharmacology, that's not drugs.'

No, that was physiology, and note that I got in to it rather negatively. Pharmacology just came along the way. I've never had any formal training in pharmacology. What I've been doing is rather like primitive painting—studying it on my own, as it were. It's been a slow process, trying to teach myself.

'But you did go into drug research.'

Yes, but the point that I'm making is I came to pharmacology accidentally. It simply was a path which appeared and I went down it.

'You never set out to find a drug?'

I haven't ever had a game plan in life. My life has been, as far as it is possible, entirely unplanned. The drug came out of ideas which arose while I was studying the cardiac circulation and the circulation through the gut. I'm always just following something, and always aiming for something which I ought to solve fairly quickly by asking simple questions.

'So, tell me how that worked with the heart.'

What I thought might take me maybe a year, turned out to take about six, but once started, then, of course, I was committed. I'm not a flibbertigibbet. Once I start something I stay with it until it's finished.

'What was so special about your approach?'

There were two elements in it. One was to think about what was happening in angina in a different way. Angina is intense heart pain caused by disease-narrowed arteries to the heart and can lead to sudden death. Instead of thinking that the problem was to find a way of increasing the supply of blood and oxygen through narrowed arteries, my experiments made me think it might be better to see if we could decrease the amount of oxygen which the heart needed. The prevailing thinking was to try to dilate the coronary arteries, whereas my project was to say, 'Forget about that, let us see if we can decrease how much blood the heart actually needs'. That was one approach. The other, which ran counter to prevailing thinking, had to do with the nerves to organs like the heart. The sympathetic nervous system adjusts our behaviour to emergency situations, to severe exercise, to fight, fright and flight. The prevailing thinking was that every reaction involving this peripheral nervous system had survival value, was always good for you. Walter Cannon had spelled out this idea in his book *The Wisdom of the Body*. The project which I set out on took the opposite view; namely that it had survival value only if you were fit. But if, in fact, you were physiologically crippled, if you had damaged coronary arteries, for example, then the reactions which in health helped you to escape from your prey or your problem, would now embarrass you. So I think these were the two elements in my approach, questioning what seemed to be accepted thinking in these two areas.

'I can see that you completely turned over conventional wisdom but that still doesn't get you to a drug to actually solve the problem. Of course, the history of drug design is a rather dismal one, it's really been by happenstance. Am I not right?'

Well, once the story is put in place, that the sympathetic nervous system puts a person with narrowed cardiac arteries at a disadvantage, then the connecting link is adrenaline. Adrenaline is secreted by the adrenal glands into the circulating blood and its cousin, noradrenaline, is secreted by the sympathetic nerves close to the muscles of the heart and blood vessels. All the reactions of the heart and circulation to stress are due to these chemicals. So the question was how to stop the heart responding to adrenaline and noradrenaline. Now, two discoveries had been made in the 1940s. First, Ahlquist had shown that adrenaline and noradrenaline stimulated the heart and raised blood pressure by different mechanisms, namely by acting on different types of receptors. Blood pressure was raised by activation of alpha receptors and the heart was stimulated to beat faster by activation of beta receptors. Drugs which could prevent noradrenaline acting on alpha receptors were known. They could be used to lower high blood pressure. Lowering blood pressure reduces blood flow through narrowed arteries and aggravates angina. Plainly, we wanted a drug to block the beta receptors in the heart without blocking the alpha receptors in the blood vessels. That is where the second discovery came in. Isoprenaline is another close relative of adrenaline which had been found to have the property that it only stimulated beta receptors. It stimulated the heart without raising blood pressure. So isoprenaline was our starting point.

'Well, it sounds simple, but you still haven't got your drug.'

No, by starting off with isoprenaline, we thought we would be able to find isoprenaline-like molecules which would still be selective for beta receptors but which would be emasculated in their ability to activate the receptors. Using isoprenaline as the template we made new molecules and assayed them on isolated tissues, pieces of heart muscle, for the action I wanted, namely the blockade of the excitatory actions of adrenaline. That took, from the beginning, about eighteen months till we found a lead molecule. Then, of course, we had to go down the road of trying to make a better compound which would not only have the activity we wanted, but which would be also active when swallowed and have a long enough duration in the body; in other words, to make it a useable drug.

'Did you test a lot of compounds?'

In the laboratory, yes, hundreds.

'Was this just a random search, once you'd got that close?'

It isn't possible to make compounds randomly. Every time the chemist thinks of a structure to make he doesn't at all think in a random fashion. He is answering a question which may have to do with the shape of the molecule, or how charged it is. The chemist is always trying to answer a chemical question.

'Is that what you were trying to do? Do you see yourself as a chemist?'

No, but I love them. I love talking to chemists. It has been one of the joys of my life. It is something that appeals to my imagination, I have no desire to be a chemist and make compounds, but I love talking about chemistry to chemists. I might not be very good at it, but I love it.

'What's so lovable about it?'

It is, in an imaginative sense, entirely open-ended and entirely pictorial. That is a vital part of my life. I daydream like mad so it's simply something which is a rich food for the imagination. You can have all these structures in your head, turning and tumbling and moving. Now we can display structures on computer screens in 3D and colour. They're just intrinsically beautiful.

'So, you daydream chemical structures?'

Yes. You make a number of assumptions. You assume that the receptor doesn't know any more about chemistry than chemists do, and you then try and pretend that you are the receptor. You imagine what would it be like if this molecule was coming out of space towards you. What would it look like, what would it do? It's grossly inefficient. Don't let me for a moment give you the idea that the process is in any way rational. It is entirely an intuitive game. It is rational only in the sense that what you imagine has to be based on sound chemical theory.

'Would you say that was your great skill?'

I don't know about great skills. I think the only difference between me and some other colleagues would be temerity. I think I ask questions of chemists and physiologists which are really quite preposterous. And I don't seem to mind being thought stupid; I don't seem to care what people think about my questions. So I think in so far as what I've done has led to things, they have all involved turning something around the other way. I think my brain automatically does this. It always is challenging the accepted view of things. So if someone says to me, 'The speed of light is constant', then I'll say, 'Well, what would happen if it wasn't?' Let's say, for example, that it was not a cosmic constant. After all we've only known about the velocity of light for less than 200 years. You must be able to ask these kinds of questions without feeling ashamed of them.

'How did your colleagues respond to your apparently very unconventional approach?'

I think in the beginning we were tolerated. I was lucky, I was given a young chemist to work with at ICI who was sort of manic in the sense that his brain whizzed about like mad. He was full of excitement. I just remember he'd come rushing in in the morning, his notebook covered with structures and he would just make it all very exciting. So he and I simply had great fun making these molecules, testing them, getting excited with each other. Once you have made the break, once you, if you like, have the lead, then the more orthodox medicinal chemists really took over from us to develop the lead into the first useful drug. At that point it wasn't flair and inspiration. It was just hard slogging; systematic manipulation.

'Did you ever imagine that the beta-blocker, propranolol, would be as successful as it's turned out to be?'

Oh absolutely not. I think you should understand that at no point have I ever professionally made claims about what I thought my work would achieve, for the simple reason I've never thought that it was going to achieve anything other than answer a question. And the same was true with the work we did later with histamine. The challenge was to make a molecule which would have the properties of allowing you to answer a physiological question. Now, if the answer had come out a different way in each case, there would have been no drugs, but there would still have been a scientific contribution.

'Was it not a conscious attempt to try and do something about ulcers that lead to cimetidine?'

No, not at all. It was a conscious attempt to take a recognized problem in the literature which seemed to me to have exactly the same shape as the one I'd just solved, and hence seemed to be soluble.

'What has given you the most pleasure, the intellectual solution, or the actual good you've done humanity?'

I admit that this recent notoriety has meant that I get some letters in from patients, and I have to admit I like them. I always reply to them straightaway; they're about the only letters I reply to straightaway. But really I don't think about it in either money or misery terms.

'You've adopted an unorthodox approach with the development of these drugs, is this unorthodox approach part of your approach to problems in life generally?'

Probably not. I think my approach to life generally is quite timid. I'm not at all comfortable in social circumstances. I can get very boisterous, and argumentative, but, in fact, it can be a nightmare simply anticipating that I have to have an ordinary conversation with somebody. Cocktail parties, for example, I find very trying.

'You have a reputation for being impatient, for being very impatient. Is that true?'

Yes, I would say that's true.

'Impatient with what?'

I suppose, in a sense, myself really. I'm usually not moving as quickly as I would like in anything, so I generally regard my performance as far below what I'd like. That makes me impatient, and I couple that with a certain amount of laziness in the sense that maybe I'm not as good as I ought to be at disciplining my life. The place of discipline in life is one which I've never really solved, because it is both a necessity and an impediment to the way I work. In other words, I have to be disciplined in my thinking in a sense, but at the same time it has to take place under circumstances where it isn't too focused. So my reading and thinking has to be fairly wide. That undisciplined element in my thinking spreads into my private life.

'And so you don't answer letters?'

Of course I do, but I am not a good correspondent.

'You've worked both in industry and universities. Are they very different?'

Oh yes. Big differences in the atmosphere and how they're managed, but I think you have to realize there are also big differences between industries. I found they all had quite different atmospheres. They are different, like people, and my usual joke is that, also like people, they suffer from common diseases; these are the diseases of size I think. I used to say that if I ever got round to retiring I was going to write a treatise on institutional pathology. I think there's some truth in the idea that management is always about wrestling to stop the organization disintegrating. Struggling with problems which people have of their identity. There are tremendous identity problems in industry. Knowing who you are and what's yours, and what's expected of you, and to contain people and their personalities within an organization. It's an unsolved, and I suspect unsolvable, problem. Every company gets some more or less unsatisfactory way of solving the problem of the individual, as against the organization.

'What are the diseases you think they actually suffer from?'

Well, I think they suffer from things which have their equivalents in cancer or inflammation. Of course, these are all conceits, to imagine that you can make an analogy between industry and organisms. But put it like this, you have in this sort of cellular approach to industry, the problems of the extent to which the individual elements are able to be in harmony or not. There are people like me, for example, who are, in a sense, bad news for industry because I was constantly rebelling. I was constantly challenging the assumptions they made. I challenged their practices, I was awful. But on the other hand, I wasn't grumbling. I was always suggesting alternatives. But there are always people like me in industry who have to be content, and it's a big managerial problem, how to deal with somebody like me since they're around in every company. It wasn't that I set out to give them a hard time. It was simply that being in there put me in an environment which wasn't conducive to what I wanted to do.

'Did it make you unhappy?'

I think in retrospect I was always sort of unhappy.

'Happier in the university?'

Yes, I was fairly happy. Yes, definitely happier in universities than in industry. No doubt about that.

'But you're clearly not an institutional person.'

I think that's the conclusion I'm about to come to. What I'm doing now expresses that. I've set up an organization here with the shape of a small company; we have articles of association and so on but it's a company limited by guarantee, which means there are no stockholders and our intentions are charitable. If we ever make any money, it will go back into research. But also one of the articles restricts the number of scientists we can have here to twenty. It's an entirely arbitrary number, but simply to emphasize the importance I think I attach to small organizations. I think twenty reflects the limit of my range. Other people could perhaps have fifty, but twenty's about my limit. You see, small is meaningless unless defined explicitly.

'You're totally free here to do exactly what you want?'

I'm free totally in the sense that the contract that I have, which brings in my money, allows me to be free; but I'm not free in the sense that I can do what I like from the point of view of being capricious. First of all I want to succeed, I very badly want to succeed. So this already restricts me to the kind of project which I set out on, and as I mentioned earlier, once you set out on a project you become a captive within it, you're no longer free. And a project, once launched, becomes a very hard taskmaster, and you can't dodge, you can't say, 'this is too difficult, I'll go and try something else'. So, I'm no longer free in that sense, I'm committed.

'It's strange that you have this need to succeed. You have been so enormously successful. You don't see it like that, though?'

Ah, um, no, I don't see it like that, no. When I'm saying I have to succeed, I have to solve the problem, and there is nothing more exciting than solving a problem. So it really is a kind of addiction.

GERALD EDELMAN
was born in 1929 and was Vincent Astor Distinguished
Professor at Rockefeller University New York, and is
now Chairman of the Department of Neurobiology at
the Scripps Research Institute, California, and Director
of the Neurosciences Institute.

CHAPTER 14

Going into the dark

❧

Gerald Edelman
Immunologist and neurobiologist

MOST scientists are content if they achieve something worthwhile in one area of research. Gerry Edelman has been remarkably successful in two different fields, and may well be on the way to the hat trick.

His first project was to unravel the structure of antibody molecules which enable the body to recognize and deal with foreign substances such as viruses and bacteria. For this work he shared a Nobel prize in 1972. He then moved into my own field, embryonic development, which is how I got to know him. Again he made a major contribution by identifying special adhesive molecules which play a key role in morphogenesis, the moulding of embryonic forms. Now he is devoting enormous time and effort to the brain, and in particular to understanding consciousness itself, and has set up a new institute near San Diego.

We now know that immunoglobulins, or antibodies, have two molecular arms which effectively capture or combine with substances they identify as foreign and neutralize them. One of these arms is much lighter than the other and is consequently known as the light, as opposed to the heavy, chain. None of this was even dreamed of thirty-odd years ago when Edelman started to wonder how antibodies worked.

It so happens, however, that in certain diseases the light chains are produced by malignant cells and can be detected in patients' urine. The diagnostic significance of these Bence Jones proteins, as they are called, have been known since the last century, but no one had any idea what they were until Edelman began to put two and two together. By using the new techniques of chemical sequencing to work out the molecular structure of the Bence Jones protein, it became possible to establish the structure of the immunoglobulins.

I wanted to know how Edelman got there and whether there was anything in his early life which made such an achievement likely.

———

Well, I don't really know the answer to that. My family background is reasonably straightforward. My father was a physician. He practised for about fifty-five years in New York City in rather impoverished neighbourhoods because he believed very much in treating the poor. And my mother was a businesswoman. By the time I had reached High School age she had retired, and we lived either in Long Island or in Queens, New York, where my father had his practice, and no, I don't think you can conclude because he was a doctor I was interested in science. In fact I started off in music—the violin—and it was only after I decided not to commit myself fully to a performance career that I moved back towards science.

'Why did you decide not to commit yourself to music?'

Well, Lewis, you might be the first to laugh at this, but I wasn't a performer; I was a fairly good musician but I concluded that I wasn't a performer. If you think of a performer as someone who transforms the occasion, the musical occasion, on the spot, and engages the audience—I realized that I didn't really have that. If you want to know an example of what I'd call a superlative performer, it would be Rubinstein. For example, his love of the music and of the whole event was so transparent and transmissible that everybody got very excited; it almost didn't make any difference what he did.

'But you recognized at that early age that you weren't a performer?'

Well, the early age was twenty-one. I began at an early age—when I was six.

'Then when did you commit yourself to science?'

I got this sort of sense that things weren't quite gelling in music, so I started to think about the whole process and the first thing I thought about was becoming a composer. I didn't have any talent, I had knowledge but no talent. Probably it's still the case, but in that case it was poignant and so I decided I must leave music. That was extremely painful I must say, although time does wonders, and when I think back on it I don't really remember the pain too well. I decided, well, maybe scientific research, because I'd read a lot about it because of my father. I thought in order to do research you had to be a doctor. That was, of course, in some people's view a dreadful decision. Even so, I decided to go into science, and I decided that the way to do that was to go to medical school. I didn't have a very well-formulated notion of what science was—when you're young you have these romantic visions without much clarity and maybe that's a good thing. I went to medical school and in the midst of medical school I had the good fortune to come upon some people like the biophysicist Britton Chance at the University of Pennsylvania, and gradually became more familiar with what science was about. But by that time it was too late, I was already almost a doctor, so I went through with that.

'I'm surprised—you went to University of Pennsylvania. Not Harvard or Yale?'

My dear Lewis, the fact is that originally I didn't get into college at all. It's a very long story. I didn't get into college and when my mother said that if I didn't go to college I would have to work, the thought was so appalling that I told my sister she must convince her Dean to accept me at Ursinus College, which is a little Dutch Reform college outside of Philadelphia. They accepted me and it was from there that I went to the University of Pennsylvania.

'Did you ever practise medicine?'

Yes I practised medicine. I have a licence to practise medicine in two states, although I'd warn anyone, even an emergency victim, to stay away. After I took my house officer's training at the Massachusetts General Hospital, which was Harvard, I was inducted into the army and my thought was I would go to work at the Walter Reed Hospital on viruses. I had it all taped, as we say. I was already in Texas when my wife learned with delight that my orders were changed to go to Paris. I was dreadfully disappointed but she was absolutely thrilled. I went to Paris but I didn't do research at all, I was at the Hôpital Américain in Paris, and there I delivered babies and saw thousands of patients.

'It's hard to see how, then, you came to discover the structure of the immunoglobulin molecule.'

Yes it is isn't it [laughs]. Well, what happened was that Detlev Bronk, who had just founded the graduate programme at the Rockefeller University, touched base with my old professor of medicine, Walter Bauer, who had always hoped that I would go back to Mass. General to finish my training. Bauer, a very fine man, a great doctor, wrote to me in Paris and said, 'I have this fellow Bronk, a friend of mine, who has this crazy programme and I know it will appeal to you but I want to appeal to your better judgement—come back to the Massachusetts General. If in fact you won't come back, I will recommend you.' I wrote back saying, 'I do hope you'll forgive me if I accept the idea of applying to the Rockefeller'. That's how I got into the Rockefeller University where I took my Ph.D.

'Still we're not at the immunoglobulin molecule?'

No, we are not. When I was in the army I was extremely bored. One day I decided I would start to read some scientific books to see if I could find something that would engage my mind. I remember reading Cohen and Edsall's book on proteins, which was a virtually untouched copy in the American library in Paris, a pure copy. Then I turned to an immunology book, it was a medical book. I remembered immunology in medical school as being sort of a little by-product of bacteriology. You didn't get much of it, about a week and a half of dogma. But this immunology book struck me in a very curious way since it was all about foreign substances, or antigens, the things that invade your body, the viruses, the bacteria, and their possible chemical shapes. But it wasn't much about antibodies, though there was a picture of an antibody. I now know it wasn't a picture of an antibody, it was a picture of an imagined antibody. It looked like a football. In those days we didn't have X-ray crystallography, we couldn't see what molecules looked like. What we did was to let them fall in ultracentrifuges, which provide a force field which makes these small molecules fall through a fluid, and then using some mathematics you could get at an imaginary model, which either looked like a medical pill or like a football. And so that got into the book and so when I looked at that football, I was sure that that wasn't at all a picture of an antibody. I couldn't imagine why they didn't spend more time on antibody molecules, they seemed to be the centre of things.

'Did you work on your own?'

Well, yes and no. I did work on my own and that was because of the peculiarity of the Rockefeller programme. In fact I was assigned by Frank Horsfall—who was a great virologist, and head of the new programme of Bronk's—to the laboratory of Henry Kunkel. But for the most part Kunkel didn't really know what I was doing, and most of what I was doing was rather physical and chemical. Fortunately there was this ultracentrifuge, this machine for creating those force fields. I promptly got a big bottle full of antibodies—they were gammaglobulin fractions—and that harks back to Cohen and Edsall whom I'd read in the library in Paris. Cohen and Edsall had been very concerned during World War II with blood substitutes, and this method of making various fractions of the blood was partly their invention. The antibodies were a gift from Lederle Laboratory and came from human placentas of all things. I ground away and looked at them in the centrifuge. I was pretty much left alone I must say.

'So, are you really saying that you solved, you really solved, the antibody structure on your own?'

Oh no I'm not saying that. Any scientist who says he solved anything on his own has quite a problem. Of course not. You remind me of that story of the man who received a prize and he stood up and he said, 'If I might quote Newton who said, "If I've seen so far, it's because I stood on the shoulders of giants", if I have seen so far, it's because I looked over the heads of pygmies.' You are not going to tempt me to say that. The fact is it didn't work that way at all. The accident of my going to medical school in innocence informed me, as you know, of the curious case of the Bence Jones protein.

Henry Bence Jones, who practised in London, was a student of Liebig. He went off to study with Liebig, and when he came back he received a letter from two Scottish practitioners, Macintyre and Watson, about a grocer who had a disease that causes softening of the bones, and turns out to be a cancer we now call multiple myeloma. They had discovered something very curious. After Liebig first discovered albumin it was noticed that if you heated urine from a sick patient, albumin would spill and the urine got cloudy. Macintyre and Watson were very astute. They had attended to this wealthy Scottish grocer, who was obviously dying, and heated his urine. It got cloudy, but then it got clear, and in a very terse description they wrote a letter to Henry Bence Jones and said, 'What is it?' That's what the letter ends with, 'What is it?' It didn't fit the albumin picture. Bence Jones then did a lot of stuff in iron retorts, you know, cooked it all up, analysed it for carbon, hydrogen, oxygen, and he came to the conclusion that it was a hydroxide of albumin. He became very famous because this particular phenomenon is almost pathognomonic, as doctors say, of the disease. If you see a patient whose urine you cook and it comes cloudy and then gets clear with further heating, he almost certainly has multiple myeloma. Well, I knew that, and tucked away in the studies I'd done on immunoglobulin was this idea that maybe we should try to find out what Bence Jones protein is.

People had been working on cancers of the blood cells, particularly the plasma cells that make immunoglobulin, and it occurred to me that there may be a connection between what I was doing and Bence Jones protein. I had found in the ultracentrifuge that if you treated the antibody material a certain way, its molecular weight dropped. The speed with which it dropped in this field was reduced as if it were lighter and had split. When I took this to all my colleagues they thought I was crazy. I've learned since that when that happens, if you can be certified that you're not crazy by some exterior agent, you've made a discovery. I mean they said, 'You're crazy, it just can't be. It's just untwisting and it isn't splitting.' Well, I stuck to it, I realize now that I did about 268 extra experiments that I needn't have done if I'd used my brain a little more. But I persevered and then I made this connection in my mind that this splitting might be due to the fact that the antibody wasn't made of one single long rope or chain but was made of several, and that maybe the Bence Jones protein was the light chain. In those days, Fred Sanger had just determined the structure of a protein, the amino-acid sequence of insulin. This molecule I was talking about had a molecular weight of 150 000 as compared to insulin's 6000. Everybody thought I was crazy because I said we're going to do what Sanger did.

'Going on to do the sequence?'

Yes. Well, there were two reasons you couldn't do the sequence, one was the size and the other was that the mixture I had was of everybody's antibodies, millions of different kinds. They didn't realize what different kinds meant then. But it occurred to me that a single patient was probably going to be making only one kind. The combination of the small size of the Bence Jones protein and its identity with what we now call the light chain, which my colleague Joe Gally and I showed in very quick order, opened the road. All this made it possible to do the sequence. Long story, but there it is.

'Did you get support from the people in the department?'

The answer is yes and no. To the extent that the Rockefeller climate then was such as to allow a young squirt like me to just fool around, the answer is yes. I consider one of the most basic aspects of scientific support is that people leave you alone, they don't hinder you with a lot of bureaucratic nonsense, or teaching, and to that extent I got the most magnificent support. I must say the Rockefeller was a great climate. To the extent that people thought I was crazy—and although they would not particularly shy away from me at lunch table, they thought that my work was pretty preposterous—I didn't. I really wasn't too concerned about that. There's a kind of innocence that helps you both ways. The innocence of not knowing how difficult it is to do a protein of molecular weight 150 000 is a good thing at the beginning, even though it might look like bad judgement to someone older. And the innocence of how scientific politics goes is another good kind of innocence I think.

'Do you think that it is a characteristic of scientific progress, that people have to think you slightly crazy, that you have to go against the stream, as it were?'

Well, not necessarily. It seems to me that any attempt to try to describe what a scientific discovery is in categorical terms is about as foolish as trying to give a theory of swimming after some fellow who doesn't know about that theory has just won the Olympics, or to give a theory of painting, or any kind of creative endeavour. You can say that there are times in the history of a subject when you are very lucky if you come along when there's a contradiction, or when there's a crisis. These small crises are terrific as long as you have enough courage or stubbornness—I don't know which it is—to push through and see what's behind it.

'What made you then begin to work on morphogenesis and to start looking for the adhesion molecule?'

Well, again it's difficult to be categorical, but I do think there are different kinds of scientists and different motives for wanting to find answers and different approaches to the whole enterprise. I felt pretty, how shall I say, assuaged, satisfied; I don't know how to put it, but I felt I had a kind of grasp of immunology when Macfarlane Burnett had put forward his clonal selection theory, and we had a reasonably a good marriage between antibody structure and molecular genetics, which is really the final chapter in the antibody story. Rod Porter and ourselves had done enough of the chemistry to give you a picture of the so-called primary structure of a complete antibody molecule, and when that happened you could see with great clarity how everything connected up in immunology. You couldn't see all the details of the field, in fact people are still working on it. But I try to define sciences as open and closed. That doesn't mean there is not an infinitude of things to discover after you have understood the basic principle, but for me once a field has that principle in hand it becomes closed. I felt that way about immunology. I didn't feel that it was all over—that's been misconstrued. Apropos of which some misguided reporter once said, 'Edelman says that when he went to the field all was darkness, when he left, all was light.' But the fact is that the field closed nicely for me personally, and I like to work in dark areas.

I like to work in open sciences if I can, I like to work on big problems, and there is no more thrilling problem than the problem of development, I needn't say that to you as a major figure in the field. There is a connection, because, if you think about it, the origin of antibody diversity and the way the immune system goes about doing its thing is really a developmental problem. But it lacks one piquant detail, which is the challenge of morphology. How are things shaped in the higher order? How do cells get together to make a nose? How do you inherit such things? By the time the molecular biologists had had their triumphs, we knew how a gene specified a protein molecule, we didn't know all the details, but it became sort of closed. But by no means did it give you any hint as to how you could inherit a nose, only a hint as how you could inherit an albumin or some other protein molecule—I thought that was really a great model problem.

'What's the pleasure that you actually get from science?'

I'm not a hedonist, I'm not an Epicurean, at least I'm not any modern version that I know of. I can only respond by indicating to you that some English philosophers, some rather eminent ones, have pointed out that pleasure might just be the absence of pain. Now, I'm not a devotee of Jeremy Bentham, and I don't have any calculus of pleasure and pain. After all, if you've been filling in the tedium of everyday existence by blundering around a lab, for a long time, and wondering how you're going to get the answer, and then something really glorious happens that you couldn't possibly have thought of, that has to be some kind of remarkable pleasure. In the sense that it's a surprise, but it's not too threatening, it is a pleasure in the same sense that you can make a baby laugh when you bring an object out of nowhere. That's the way I'd describe it, I had the baby laugh. A thing comes out at you as a realization of the extraordinary complexity, power and diversity of evolution. For the rest it would be, I think, mawkish and foolish for me to say that I bound into the laboratory every day and say, 'Let's go for pleasure boys.' The fact is it's like everything else in this life, you grind along, you probably don't fully know why you're doing what you're doing. But a good part of expectation is habit isn't it? Breaking through, getting various insights is certainly one of the most beautiful aspects of scientific life.

'Is that what drives you then?'

I can't answer the question because when you say 'Is that what drives you?' I immediately go to the wall. I try to think of what are the larger reasons; this, that and the other thing. I can't possibly say. It's extremely difficult to say that, well, I'm going into the lab again tomorrow because maybe I'll have another epiphany. Experience says forget it. So that, no, that isn't what drives me. Curiosity drives me. I believe that there is a group of scientists who are kind of voyeurs, they have the splendid feeling, almost a lustful feeling, of excitement when a secret of nature is revealed. I don't know how that fits in, and we're not going to get into psychiatric matters are we, because that would embarrass you, so I won't do it. I do think that there's a group of scientists who are driven very largely by curiosity and I would certainly place myself in that group. I'd really like to know how things work, and, if you really want to know, I will go against my better judgement and say what I'd really like to know above all, before I go into eclipse and oblivion, is how we can be aware of things, conscious, and how the whole world fits together. Now, of course, as a scientist you realize this is hopelessly ambitious.

'Does that encourage you?'

Well, it doesn't discourage me.

MICHAEL BERRIDGE
was born in 1938 and is Professor at the Laboratory
of Molecular Signalling, The Babraham Institute,
University of Cambridge.

CHAPTER 15

Hole in one

❦

Michael Berridge
Cell biologist

MICHAEL Berridge is probably Britain's most quoted scientist. The winner of a whole batch of international awards, and widely tipped for a Nobel prize, he remains virtually unknown outside the scientific community. This is even more surprising given that his discovery is central to understanding many aspects of biology. By unravelling a mechanism by which cells respond to the signals they receive from hormones, the chemical messengers which enable different parts of an organism to communicate with each other, Berridge has cast light on processes as diverse as growth, and why lithium acts to alleviate manic depression. He began his research in Cambridge, doing his Ph.D. under the distinguished insect physiologist, Sir Vincent Wigglesworth. But he was born and grew up in Rhodesia, which is now called Zimbabwe. What set him off in the direction of a scientific career?

———

I've always had a great interest in the natural world, which probably began before I went to university. Even as a schoolboy I was absolutely fascinated by wild animals, and in fact one of my initial objectives was to become a game warden, surprising as that might be. I intended to go to university and study big game ecology with the intention of remaining in Rhodesia and working within the game department. But when I got to university my interest changed. In fact it was quite a dramatic change because my interests switched from big game to insects. It was mainly due to one of the lecturers at the university, particularly Dr Ina Bursell, who gave such fascinating lectures on the insects. He developed my fascination for insects, and then I was attracted by the prospects of going on and doing research on them.

'Were you already ambitious to do research at this early stage?'

No, I wasn't, because it actually took some time for me to realize that there was a research aspect to university life. The university I went to in Harare was, in fact, a very new university. It had only been in existence for a year before I arrived there, so there was absolutely no tradition of long-term research.

'So how did you get to come to Cambridge?'

Well, I wanted to do insect physiology and at that stage there was no Ph.D.

programme in Rhodesia. Dr Ina Bursell had been a student of Sir Vincent Wigglesworth at Cambridge. He suggested that I might apply to go and work with Sir Vincent and arranged for me to go to Cambridge to do research. Coming to Cambridge from Rhodesia was a tremendous change. In Rhodesia the total number of students was about 200, covering all of the disciplines, whereas Cambridge had an enormous number of students and lots of different disciplines. It was a completely alien world to me.

'What did you feel about it?'

Well, I can remember sitting in my room in Caius College. I'd been there for about four weeks and had achieved practically nothing. I can remember wondering about why on earth I was here. I really felt that I just couldn't understand why I'd left home, travelled all this way. Here I was sitting in this magnificent room, but feeling that I was totally incapable of actually going on. I had a real sense of despair at that moment. And this was heightened by the fact that the postdoctoral people within the Zoology Department seemed to take a great delight in telling new research students that Sir Vincent always took on three students each year, but that one of them always dropped out. The other two students who had started with me were both Cambridge people, so they'd both got a running start. I really felt that I was the one destined to drop by the wayside. But I struggled on and after about six months, indeed one of them did pack in, so I then felt a lot happier about my future prospects.

'Were you aware of your colonial background?'

I suppose I was to a large extent. One's reminded of it constantly because of one's accent—so that often comes up—but I suppose in a way a colonial background is not such a bad way of starting in academic life. In many respects it's a very disciplined sort of upbringing, so you are, I think, given the right kind of credentials before you start.

'In what way?'

One of the aspects is that you're taught to try and succeed, and I guess this comes a lot through the sporting ethos. When I was in deep despair I did feel that I had to soldier on and try and make a success of it rather than run away from it.

'You're famous, very famous, for your work on cell signalling. What is it exactly that you discovered?'

I suppose the best way of trying to describe it is to say that I've discovered the way in which hormones regulate cellular activity. One can think of a cell very much as a closed box surrounded by a cell membrane. Hormones are signals which arrive at the cell surface and in most cases they never invade the privacy of the cell. They interact with receptors on the outside, and somehow the information from the hormone at the surface has to be translated into a message that the cell understands. It turns out that there are relatively few messengers that the cells understand, and one of these turns out to be the messenger inositol trisphosphate, which we discovered and worked out just exactly what its function is.

'So, is inositol trisphosphate in effect a second messenger that responds to the hormone arriving at the cell surface?'

That's right. The idea is that the hormone is essentially the first messenger, but then the information has to be transduced into a form that the cell can understand. Inositol trisphosphate is the second messenger which understands and activates particular cell functions.

'Were your ideas accepted quite quickly?'

We went through a period when there was quite a lot of opposition to the idea that we had found a second messenger. We had quite a lot of hard work to do to convince people. I think we were helped a lot by the fact that we were very open initially about the discovery. My colleague, Robin Irvine, was also very open in terms of dispensing our molecule to everybody who requested it. In this way many people tested the molecule very quickly and found that, indeed, it did work as a messenger. So, after the initial reluctance of some people to accept it, there was then fairly general acceptance that it did function as a messenger.

'You speak about this openness as if it's a rather rare phenomenon.'

I think that's true. Many scientists when they discover a new molecule tend to be rather possessive about it, and they try and hold on to it for as long as possible. I can remember discussing with Robin what approach we should try and adopt. We arrived very quickly at the idea that we should be as open as possible. And that absolutely paid off.

'There is in fact an image of current science as being intensely competitive and rather nasty. That doesn't seem to have been your experience.'

That's not been my experience at all within the cell-signalling field. It's been an absolute joy to work in it because relations between all the different laboratories have really been excellent. I think this is quite unusual because there are fields where, as you say, it really is quite nasty.

'You describe a lot of your work in quite emotional language—depression, elation—is that how it's been?'

Well, it's certainly been a rollercoaster, certainly the last ten years.

'A rollercoaster in the sense of excitement or ups and downs?'

I think excitement, especially as one became aware that inositol trisphosphate is a very important molecule. As it functions in so many different systems one really felt that one had discovered something extremely important.

'But were there periods of depression and gloom mixed in?'

Doing research there are always long periods of gloom because as you then start trying to take a discovery further, you hit difficult periods. In general I like to think of these more as fallow periods, and although you often are very depressed during these periods when nothing much is happening, I think it's important to realize that you've got to work through these. To my mind, these are the most important periods in a scientist's life. It's easy when things are going well, everybody can cope with that. The real difficulty is when you do come to a period where nothing much is working and it's what you do during these periods which really determines what sort of success you're going to have in the future.

'And what does one do in those periods?'

In the period before we really got on to the inositol trisphosphate story, I remember starting to measure inositol trisphosphate inside cells. But after a while I found we were simply repeating experiments, getting the same data, but not really moving forward. We entered into quite a prolonged fallow period. One way I like to try and get out of these is to read; that is, to go into the library and simply immerse myself in the literature to try and find a new angle, a new approach to the problem. It was, in fact, during that fallow period that I came across the work that people had done on lithium, and this was perhaps the single most important development that occurred to actually bring out the importance of inositol trisphosphate.

'Is this lithium in relation to the treatment of psychiatric illness, or a general effect of lithium?'

This was specifically the work that people had done on lithium with regard to its role in controlling manic depressive illness. Workers in St. Louis, particularly Allison, had made some preliminary measurements showing that lithium did indeed influence the level of inositol in the brain of rats. Nobody in the field had picked up on these early papers. When I read about it the significance of that work hit me like a ton of bricks because I'd already lived through the discovery of another messenger, called cyclic AMP, and one of the most important aspects was having molecules or drugs which could be used to interfere with that messenger. And here buried in the literature was a drug, lithium, which looked as if it interfered with our messenger molecule. I immediately started doing experiments using lithium and that led directly into the discovery of inositrol trisphosphate as a messenger.

'You do a lot of your experiments yourself.'

Right.

'Do you think that's important?'

I think it's extremely important to spend as much time as one can working at the bench. I don't do as much of it now as I would like to. It's very important to actually sit and do the experiments because so many ideas come to you during the simple process of carrying out the experiments. It's very difficult to actually think of new ideas and new concepts, without really understanding the experimentation behind it all. Also, good experiments should, now and then, give you the unexpected result. All the time you're carrying out the experiment you're thinking, why am I doing this, what am I trying to show through these experiments? Nine times out of ten you get the result which you're hoping for, but then very occasionally you get the unexpected. So you repeat the experiment and if the same thing happens again, then you begin to get the feeling that there's something unusual here. Then one has to design new experiments to have new ideas. It is the constant interplay between intellectual aspects and the physical aspects of experiments that I find really enjoyable.

'Do you enjoy the life style of science? I mean the meetings, the colleagues, the writing papers and so forth?'

Yes, I think that is perhaps one of the most enjoyable aspects of being in science. If I was hidden away in a soundproof room doing experiments and never being able to describe them to anybody, it would take away all the interest. I remember listening to a programme on Mendel. Apparently when he'd done his experiments and worked out some of the basic laws of genetics, his colleagues were so stupid that they couldn't appreciate what he'd actually done. I can imagine that this must have been enormously frustrating for Mendel, to realize that he'd discovered something really important, and yet he couldn't tell anybody about it. For me, most of the pleasurable moments in the lab are when you have the informal discussions, going over the new results and realizing how exciting they are and then planning future experiments. That's really what makes a scientist's life worthwhile.

'Do you spend most of your time thinking about science?'

Almost all my waking hours are spent thinking about science. It's not a nine to five occupation. To be a successful scientist you really have to worry away at things all the time and even hope, when you're asleep, that your subconscious is continuing to worry away at some of the problems. It's rather like being a little terrier and worrying at things until you can get a solution. It is a completely all-engrossing, full-time occupation.

'Does anything else give you the sort of excitement, and depression, that science gives you?'

Perhaps the closest one can come to doing science without actually doing it is playing golf.

'I don't understand.'

This is not an original concept, but it's a concept which has been put forward by my golfing friend, Peter Lawrence. Playing golf, the sort of golf we play, you usually have about ninety shots, and out of those perhaps one or two are really good. And doing science is much like playing golf. Almost every shot in golf, the way we play it, is a total abject failure, but very occasionally you sink a long putt, or hit a good drive, and this gives you the motivation to go on. There's a very interesting analogy between the two pursuits.

'What is it, then, about science that you like? I can see what you like about golf, it's the good shots, what do you like about science?'

Well, I think it's the same as playing golf, it's the challenge to try and go out and master this difficult pursuit. I find science is extremely difficult because you're up against nature and I like to think of it very much like a battle. You're like a General marshalling your forces to try and unlock some of the secrets. I see nature guarding these secrets very carefully, and you have to use all your skill and wits in trying to get at those secrets. I really enjoy the tremendous intellectual challenge and the occasional exhilaration of uncovering some of nature's secrets.

'What sort of skills do you think you require in this battle? What does it mean to be gifted with respect to research?'

It's a combination of things. To be a successful scientist you need to be able to make connections. If I was thinking of any single gift that one needs it's this ability to make connections between a lot of disparate facts. That's what I enjoy—particularly going into libraries, sitting down and reading, and trying to collate different pieces of information. The gifted scientists that I meet and enjoy talking to do have this facility. They have a very broad view of what's going on and they're able to make connections between different ideas, different disciplines.

'You are one of the most quoted scientists in the world, yet as far as I know this is the first interview you've had for the radio and you've not even done television, why do you think you've been so neglected?'

Well, that's a hard question to answer. You may have to go and ask the people at the BBC, or the newspapers, why British scientists tend to be so ignored. I do remember when I won the Louis Jeantet prize it was hardly reported in the English press, whereas the winners of the prize in France received enormous press. I think the English do not like success by scientists, or anybody else. We tend to denigrate successful people. It is a puzzle why scientific success is so neglected here. In the United States scientists tend to get a lot more press than they do in this country.

'Now, to ask you something more difficult. You are widely tipped in the literature publicly as a very strong candidate for a Nobel prize. How does that affect you?'

Many people make that very statement about lots of scientists. When they refer to me I find it a very uncomfortable question to have to deal with. When the suggestion first arose, it did create a certain amount of excitement. Of course one's excited, naturally, by the prospect that one might win the Nobel prize, but after a while you begin to realize that this is something that could become an obsession. I've now passed through that phase and I don't worry about it as much as I did before. At the present time it has very little influence on me.

'Do you ever have a sense of awe that your discovery is so important?'

I certainly do, yes. I never really thought I would ever make any sort of discovery of this kind. It's not something that comes to everybody. It's not so much a sense of awe, but rather the feeling that this is such an unexpected thing to have happened to me [laughs].

ELWYN SIMONS
was born in 1930 and is Professor of Anthropology and
Anatomy at Duke University, North Carolina.

CHAPTER 16

Ways of seeing

∽∾∾

Elwyn Simons
Palaeontologist

THE name most people connect with the search for human origins is Leakey. Between them members of the Leakey family have found many of the key fossils which tell the story of our evolution from ape-like hominids to *Homo sapiens*. But what about the earlier chapters in the saga? What kind of primates came before the relatively advanced creatures which gave rise to ourselves?

The man to answer these questions is Elwyn Simons. Master fossil hunter, conservationist and widely regarded as the founder of primate palaeontology, he is concerned with all the stages of human ancestry, but particularly with the fossil remains of the small, monkey-like creatures which lived in the Oligocene and Eocene period between 40 and 30 million years ago, in what is now the Fayum desert of Egypt. The most famous of these is *Aegyptopithecus*, which is probably the first of the line which branched off and eventually led to human beings. Other Eocene primates gave rise to the lorises and lemurs which Simons studies in the forests of Madagascar and at the Primate Center at Duke University in Durham, North Carolina.

Looking for tiny fragments of fossil bone in remote and uncomfortable places presents a very different image of modern palaeontology than the high-tech end of evolutionary research where molecular biologists attempt to reconstruct the past by delving into the DNA of living primates.

What are the skills of a palaeontologist who works in the field? How was Simons drawn into studying our pre-human ancestors?

———

It started very young. I started drawing pictures of animals. Some people draw pictures of cars or horses, I just did various animals and kept pets, I couldn't have enough pets.

'What did you have?'

Well, nothing too dramatic, but I had one thing that's called a hog-nosed snake—it sort of mimics a rattle snake by hissing and flattening itself out—and it got loose in the garage. My mother didn't appreciate that [laughs].

'But did your mother encourage you in this interest?'

Yes, both of my parents were interested in biology and I think my interest started first in this love of animals, but there were also intellectual issues I got interested in. I have always been interested in history, in the past, interested in my own family history, and some people just aren't. I do genealogy as a hobby, and you meet people that don't know anything about their family and are glad they don't. I wasn't that way. It's kind of inexplicable. My great-grandfather actually came to America from England, and I remember when I was, I think, about five, my dad telling how his grandfather had sailed across the Atlantic in a ship that took six weeks, and that was intriguing to me, you know, to learn a thing like that which another child might have forgotten. And then extending from that, maybe, the history of life interested me. I wanted to know where we came from, why we're here, the meaning of existence and life, and to do that I soon got on to studying human fossils.

'At a very early age?'

Oh yes.

'How old?'

Well, I've got a book called *The Earth for Sam*, which was given to me on my eighth birthday, and that has some introductory information about human evolution, so I must have been reading it from then on.

'So, your career was set from an early age, in other words?'

Well, I think it could have gone in many directions, but it probably would have been in some area of science. I had another interest, though, which was art and drawing, and I got an art scholarship, which the children in public schools in Houston, Texas, where I grew up, were eligible to compete for, art scholarships at the local art museum. I think I got that when I was nine and it meant nothing very dramatic, but we had free art classes every Saturday morning, and could stay over in the afternoon if we wanted to. And we had some quite good teachers there that were on the staff at the Fine Arts Museum.

'And you were tempted to go into the arts?'

Yes, I rather thought I could be a portrait painter or something, but I reached maturity just at the time when colour photography became so impressive, and also all of my teachers said that modern art really was non-representational, it should be like Jackson Pollock, you know, just paint dashed around in various directions. So I thought, well, I don't want to do this non-representational art, so I went on to use my ability to draw in my science, instead; drawing pictures of fossils and other things.

'So when did you actually begin to hunt for fossils?'

Well, it was a hobby. I had a chest-of-drawers in my room with a lot of marine fossils in it. My family come from eastern Kansas where there are Pennsylvanian marine rocks, and crinoid stems are quite common there, as are brachiopods and other marine fossils.

'Those are invertebrate marine fossils?'

Right, but they're all over the ground out there, and out in the fields, so my early collections were made from these marine sites in Kansas.

'But serious fossil hunting?'

Ah, serious fossil hunting. Well, I guess that started when I went to graduate school. I went to Princeton to work with early mammal and primate fossils, but the professor I worked under there, Glenn Jepsen, was an expert on the mammal fossils of the Palaeocene of Wyoming, an epoch that runs from just immediately after the time of the dinosaurs, for about 10 or 15 million years. He was sort of a world expert on that time period, so that first summer after I graduated from college, which would have been the summer of '54, I went out to Wyoming and collected for him.

'Was that an exciting experience?'

It's fun to find fossils because you never know what you're going to find and there's always a chance that you'll find something quite unusual, and that kind of excitement makes it sort of like a treasure hunt. Just hunting for a fossil primate, you often see the jaw first, but you can usually not see it too well because it will be covered partly by matrix, so you get these tantalizing glimpses. And that's happened several times. It's as if you know the difference between fool's gold and real gold, or could tell an uncut diamond from a piece of quartz, so when you find it, something that you know is good and new, it's quite exhilarating.

'Can you give an example of that?'

Well, there are three skulls of *Aegyptopithecus* known, and I found the last and probably the least complete one. This occurred after many, many years of working in Egypt, twenty-five years, when other people found the other two. So I finally found one, and I did a little chuckling to myself because I was alone when I found it, and there was a footprint—somebody's tennis shoe—about ten inches from the skull, so I knew that somebody else had been there and didn't see it, you see. You get a kind of a thrill out of something like that for a variety of reasons.

'I don't have a sense of what it means to go on a fossil hunt, how one plans it, and what one's going to look for, or even how long one goes for.'

Yes, you have to start with some sort of an objective, which probably means the time period that you're trying to work on, and then you can either pick a place where other people have been, or you try to find out, by acquiring maps and recent geological publications, where rocks of that age might occur that haven't been explored. So when you find a place that looks like a good prospect for exploration you go there, and then there are many ways to explore it. The most conventional way for locating fossils out in the open is what is called surface prospecting. You just wander around looking for bones. That's not too difficult to do, but still some people have a far better eye for it than others. Some people can see a little jaw on the ground, perhaps a fragment of a jaw, only a centimetre long with four of five teeth in it, each of which is smaller than a pinhead. Other people don't notice things like that when they're walking along.

'What's the alternative?'

Crawling [laughs]. Professor Marsh, whose job I held at Yale about 100 years after him, was a great organizer of expeditions in the early American West, and in the early 1870s he took out groups of Yale graduates in the year they graduated to help collect fossils, and he told them that they all had to crawl, and that they had to keep their eyes six or seven inches off the ground.

'Do you do that yourself? I mean do you get down there and really, not exactly crawl, but minutely examine the terrain?'

Right, well, if you find a spot where there are lots of pieces weathering out you sit down and then you look at everything. You pile up every little particle, every little lump of rock, and look at it to see if there's any bone protruding. I always said in the early years when I worked out in Wyoming that the ideal collector would be a big, fat man that didn't want to walk very far and sat down a lot [laughs] . . . because every time people sit down they find things. Even when they're tired, a person will sit down just because they're tired, and by gosh there between their feet is some jaw or something, a tooth that they wouldn't have found by just walking by.

'Do you think there's a real skill in knowing how to hunt?'

Yes. I think that the best collectors are like great pianists, or great painters, they really have talents that are almost inexplicable, but also it usually takes a lot of time to get really good.

'You have no idea what the skill is though?'

Well, I think it's a form of seeing. It's seeing order in a random background. For instance, in the Egyptian desert where we hunt fossils, the desert surface is all covered with stones of all sorts and colours that have survived from wind erosion. It's called desert pavement, or *serir* in Arabic, and this *serir* is a very jumbled mass of lumps of rock of all different colours and if there's a bone with a tooth in it in that background, it's not easy to see that in the pattern. I guess it's kind of comparable to some people who, if they're given a book in which some word occurs only once, can flip through and find it. There are people who can scan pages very rapidly and find a word like that, in other words, something that's very rare compared to the surface of print in all those pages.

'What about the physical discomforts?'

Well, if you like camping out there aren't too many discomforts. I don't know, some people don't like being out where it's very quiet and not too many people to talk to, or there's no radio or TV. I've known a few people complain of that. Of course I wouldn't like the field so much if I were like that.

'Do you go with other people or do you hunt alone?'

No, I'm always in groups, and just a few times on my expeditions I've been left alone in camp for one reason or another, and those are the only times I can remember in my life when I've actually been alone. I've spent my whole life surrounded by people.

'Did you enjoy being alone?'

Yes. It was quite interesting. I enjoy the different environments that you get in the field. You can go to the tops of mountains, I've been higher than 11 000 feet collecting Eocene mammals that are in sediments that have been uplifted but were deposited near sea-level. Most sites are just banks of clay where you're digging out bones, probably in a rather arid environment, but some of the other collecting I've done, such as catching lemurs for the Primate Centere, has taken me into jungles. On this last trip to Madagascar I just came back from in December, I had to live in a village that was sort of the way England or Europe was in the twelfth or thirteenth century, and I loved it. I enjoy all that, I enjoy being in those different places.

'Are you usually successful?'

Yes. I have luck, you could say . . . Dr Louis Leakey always called it Leakey's Luck. But I think it's just the natural result of persistence and knowing that you go to the right place. When anyone's persistent at something they have to be doing it because they like to do it, and we all know that what particular people like differs, but I always wanted to find fossils so it's easy for me to be persistent at it. I know when I was going to graduate school one of my friends said, 'well, what do you want to do?' And I said, 'I want to have a career doing something that always seems like play.' But it takes hard work. At one of our best sites in the Fayum we have to sit there in quarry for long, long hours and find many, many things that aren't very interesting to most people. That takes persistence.

'Is it a competitive area, fossil hunting?'

Well, it has been in times and places, and people tend to put their . . . I guess I don't know whether you can do it with a fossil or not, but they put their best foot forward for their finds. They endeavour to show that what they have found may be more important than what others have found. At least, I've seen that sort of syndrome in some of my colleagues. So people are competing to find things that they can represent to the world as being important. What is important in a fossil? Well, something that's a direct ancestor compared to something that belongs to an obscure side-branch.

'That's just what I want to come on to. I mean you spend all this time collecting them—why?'

Well, for me, in the narrowest aspect of my profession, I've been trying to elucidate the history of primates leading up to humans. And as you go further and further back there are some gaps. We know humans are derived ultimately from some kind of extinct apes, so then you can study apes and you get much further back in time. Then you have monkey-like creatures, and finally only primates that look like lemurs, or bush babies. I've studied all of these different phases of the history of primates leading up to humans, and many of my colleagues don't do that—they work either only with palaeo-hominids, or maybe they work only with Palaeocene or Eocene archaic primates. Or there are some people that specialize in the apes of the middle Tertiary and aren't very interested in earlier or later things. One or two people I know like to do only fossil monkeys. But I've gotten involved in all of these different periods at various times.

'But do you really need more fossils?'

Well, to tell the detailed story of the evolution of primates and humans, yes. But not to prove that it happened or anything like that. I keep looking for more and more details, that's why I need them.

'Why do you need these details?'

Well, fossils tend to be very limited in scope. The commonest kind of fossil is a lower jaw of a mammal.

'Why is that, by the way?'

Jaws don't break very easily and they don't have a lot of tasty meat around them so they're not ground up by carnivores such as hyenas and other predators that might crush them all to pieces. Curiously for the limb skeleton, the humerus, the upper arm bone, is the most resistant, for reasons that are less clear.

'Sorry, I took you off this issue of details, of why you needed these further details?'

Well, because you find the jaws and teeth, maybe sometimes just teeth, that are scattered around, not even in very complete pieces of jaws. The detailed story of what happened in evolution requires finding the complete skulls and skeletons ultimately, and if you take a group like primates, the ancestral primates in general, there are dozens and dozens of genera and species that have been found and described, but there are only a handful that are represented by a complete skeleton. I mean we call something that's 90 per cent complete a complete skeleton.

'And when you do find the whole skeleton, does it usually fit in with what you thought the whole skeleton looked like?'

No. What I always tell my students is that fossils are never what you expected. It's not possible, no matter how we connive and plan and try to predict what the past was like, it's never what we thought it was going to be.

'When you bring the material back, then you have to work, do you yourself do the description, or do you give it out to other people?'

Well, I work with teams of people, but with some of the more complete specimens of primates I like to do the initial morphological description myself. I've been a co-author on so many papers and I don't like to get completely lost in a welter of colleagues, so it's, I think, a little bit mandatory that by doing some solely authored papers, I remind people that I'm not dead, or I'm not having my students write them all. You know, I've been writing papers for about thirty years so people could suppose I'm already played out—and I hope I won't be for some considerable time because I've a lot more to write about.

'Do you use many modern techniques for your studies?'

We use modern equipment sometimes, in the lab. When we're studying very small things we may use electron microscopes, and work related to that uses modern biochemical techniques to determine molecular distance data. But in the field the human eye is still something very important. We have to have the best means for getting there, of course, so then our equipment ceases being a microscope and turns into some kind of excellent field vehicle. It's our scientific instrument because it takes us to where the fossil evidence is

'You've been described as the father of primate palaeontology. What does that actually mean?'

I'm not sure. I got started earlier than most people and I certainly had a lot of students that were interested in the same thing and this started a growth of interest, but I think it would have happened somehow with other personalities if I hadn't existed.

'Was it that primates had been neglected?'

Well, there were some funny things, you know. I remember George Gaylord Simpson used to kid and josh around and say there were more people studying hominid fossils than there were hominind fossils, and it seemed like that in the thirties and forties, I'm sure.

'But also you have a passion for living primates?'

That's right. Well, I was trained as part of a wave of people studying fossils right after World War II. Palaeontology had started out as something very dry—dry as dust, as some people said—with people studying fossils as evidence for the age of the rocks. But my group after the war was more interested in how they lived and moved, what made them tick, what were they really like—and to be able to infer that you have to extrapolate from what we know from living relatives. So, that almost demanded that we studied the living primitive primates in order to see what were the locomotor systems, what were the life styles that we could infer and so on, and a lot of interesting things have come out. For instance, we found out that the Oligocene primates of the Fayum have jaws and faces and teeth that are different in size between males and females. Now from studying living primates we know that that normally happens only in primates that have formed large social groups. You would think in looking at fossils that have been dead for 34 million years you couldn't possibly know what kind of social group they had, but by extrapolating back and forth from the fossil to the living you can infer a lot of interesting things.

'It is a field where, as you say, new ideas are possible. Is that one of the things that attracts you to the field?'

It does because it's a subject which allows for individual interpretation. It's not cut and dried like the periodic table of the elements. Each person may somehow sooner or later chance on a slightly new idea in this field. Some people have said to me, 'Oh, it's all so controversial and nobody can seem to agree', and I say, 'That's just why it's interesting'. Probably it is because it's a relatively new subject, or it hasn't been able to mature because our information is fragmentary. We don't get all the details that we need to have everything sewn up, cut-and-dried and agreed on by all, and that makes it exciting to be in the swim of that.

'Are you still going to go and hunt for more fossils?'

Yep. I'm not slowing down, I'm actually very thankful for that. I've had excellent health to be able to carry out field work. Sir Wilfred Le Gros Clark in the 1950s always used to say to me, 'Oh, my dear Elwyn I'm so happy for you, you'll live much longer than I and you will be part of all those discoveries in the future which will make us really understand how humans evolved.' And that's true, I am part of many things that have happened since then—but we're still not there with final answers.

MURRAY GELL-MANN
*was born in 1929 and is R. A. Millikan Professor
of Theoretical Physics at the California Insitute of
Technology, California.*

CHAPTER 17

Three quarks for muster mark

⌇⌇⌇

Murray Gell-Mann
Theoretical physicist

PARTICLE physics is, to most of us, a mysterious and unimaginable world with vocabulary to match. The behaviour of nuclear particles is accounted for by their being composed of basic building blocks called quarks, which cannot be directly detected but, nevertheless, possess characteristics dubbed colour and flavour, where the flavours have names like strangeness and charm.

At least some of the responsibility for this exotic lexicon has to be laid at the door of Murray Gell-Mann. It was he who conceived and christened the concept of strangeness, and proposed that the protons and neutrons which make up the atomic nucleus, themselves consist of quarks, a term taken from James Joyce's novel, *Finnegans Wake*. For some of these insights he was awarded the Nobel Prize for physics in 1969, by which time many already regarded him as one of the few possible successors to Einstein. When, I wondered, had he realized that he had abilities that were out of the ordinary?

Well, I knew that I was further advanced in school than most people my age, but I didn't have any clear idea of why that was so. For many years I attributed it to just having started earlier. My brother taught me to read from a biscuit box when I was three, and I had thought that probably if other people were taught to read when they were three, they would also be several years ahead in school. Nowadays I have somewhat different ideas about the relative influence of nurture and nature, but when I was a child I was convinced that it was mostly environment that counted. Once I had children of my own, it became clear to me that much is already determined at birth, even though a great deal is controlled by the circumstances of the first few years of life.

'When did you recognize that you were going to be a scientist?'

When I was a child I was very interested in such subjects as archaeology and linguistics, and natural history. I knew a certain amount of mathematics for a child, but I never intended to do physical science. When I was applying for admission to Yale University, however, I had a conversation with my father. He was appalled at the idea that in answer to the question 'What will be your principal subject if you are admitted to Yale?' I might put down linguistics or archaeology.

He said, 'You'll starve'. He was profoundly influenced by the Depression, and was genuinely afraid that if I took up something like that I wouldn't make any money. He kept citing the example of Schliemann, who had made a fortune in business before, at the age of forty or so, he started to indulge his passion for archaeology. I asked my father what he thought I ought to fill in on the form, and he said, 'Engineering'. To which I replied, 'Well, I'd rather starve.'

'Would you really rather have starved than do engineering?'

I thought I wouldn't be very good at it, and that I wouldn't enjoy it. I knew that if I designed something it would fall down or break.

'So, you put down physics?'

I asked my father what else he would suggest, since he didn't seem to approve of archaeology or linguistics, and I wouldn't consider engineering, and he said, 'Physics. That would be a good compromise.' Well, I'd taken a course called physics in High School and it was the dullest course I'd ever had. It was also the only one in which I had gotten a bad grade, close to failure actually. We had to memorize the seven kinds of simple machine: the screw, the pulley, the inclined plane and so on. And we learned about electricity, magnetism, wave motion, sound, etc., without any hint that there might be some connection among those subjects. It had been a terrible course, and I said to my father, 'You don't really want me to study that do you? Physics seems to be a hideous subject.' 'Well', he replied 'it gets better later on.' He had an amateur's interest in mathematics, physics and astronomy, and frequently—perhaps to get away from his family—he would lock himself in his little room to study certain aspects of those subjects. He insisted that physics would get more interesting when I started to study quantum mechanics and relativity. At that point in the conversation I figured that I might as well write down 'physics', because if and when I was admitted to Yale with a generous enough scholarship to permit me to attend, I could perfectly well change to a different major subject. But then, when I got to New Haven, I was too lazy to switch and, indeed, it turned out that quantum mechanics and relativity were fascinating, just as my father had promised, so I just went on to become a physicist.

'And when did you realize that you really would be able to make a contribution to physics? Did you decide to do research as soon as you became interested?'

Oh yes. But, it was not clear for a few years whether I was making a significant contribution or not. That was something about which I was very confused.

'Confused in what way?'

My problem was that I didn't know, when I thought of something, whether it was an interesting new idea or some trivial notion that was already familiar to many physicists. It was very difficult for me to take my thoughts seriously enough to make a careful literature search to see if these particular ideas had already been put forward, and if not, then to write them up carefully and submit them to some journal. So I never really knew whether what I was doing was important or not. Looking back I see that, even when I was a student, I did think of a few things that were not generally known and could have been research publications, but I was unaware of it at the time.

'You were curiously unambitious it seems to me.'

Well, no, perhaps it was, in a strange way, a consequence of hoping that I might make some gigantic contribution about which there would be no doubt that it was important.

'So, did you feel yourself at this early stage to be particularly gifted?'

I suppose I wondered if I might be, but I didn't really know. Good students aren't always capable of creative work.

'You were very young, nevertheless, when you made a very major contribution.'

Theoretical physicists make their contributions when they are young. In fields that are data-rich, or that require a vast amount of experience, the tendency is rather for contributions to be made later in life. But in mathematics and in theoretical physics, it frequently happens that some of the best work is done early. Why that is true is an interesting question, to which I don't know the answer, but one can speculate about two very simple explanations. One is that as you gets older, you have, or feel you have, more and more of a stake in existing paradigms. You may even have invented them yourself. In any case you would be more reluctant to challenge them. Another, even more obvious explanation, is that people get more to do as they grow older. They have more responsibilities at home and at the office, and so less time to think. In a data-rich field, or one in which experience is really important for some other reason, these effects could easily be overcompensated by the advantages of maturity, but that would probably not occur in mathematics and theoretical physics. In those fields freshness and concentration may outweigh maturity.

'What would you say your skills were?'

I don't know exactly, but I do have a tendency to take a very broad view; to try to see the big picture and connect a lot of things together. I particularly enjoy seeing widespread connections but, of course, one has to watch out because many apparent connections are fake. In fact lots of crackpot ideas consist precisely of postulating widespread connections—conspiracy theories, for example. But still, working on the big picture is something I enjoy, and I think it's helped me at times. Of course, I also get a kick out of using mathematics in the service of science.

'Is being meticulous important to you?'

Yes, it is. I'm an unusual person because I like vague, general, wide-ranging notions and at the same time I tend to be meticulous and analytic.

'In your career have there been times when you've been beset by doubts?'

Oh, I have always lived with a great deal of doubt. No matter what I've come up with, I've doubted it and worried about it and been terrified that it was wrong or trivial. I've never experienced unalloyed, unadulterated joy in any thought or discovery. It's always been tempered by uncertainty and worry.

'And that continues to be the case?'

Oh yes.

'That's extraordinary for someone who's been so successful.'

It's true, though.

'But do you see yourself as having been successful?'

Well, I have done some useful things, certainly, and I'm very proud of them. But at the time the things that weren't being explained, along with the apparent difficulties and the possibility of being wrong, always loomed very large for me. I don't say that's a healthy mental attitude, I'm just being honest and saying that it was, in fact, how I felt.

'Did you find being wrong painful?'

Oh yes. I really hate to be wrong. I don't actually deny it when it happens, but I don't like it at all.

'Do you think that your very doubts have contributed to your success?'

On balance, I think probably not. I have often worried so much about things that I've delayed putting forward useful ideas for a very long time. I think the extent to which I've worried about things has been a little bit pathological. Certainly one should be sceptical of one's own ideas, and careful, but I probably carried it too far.

'How do you actually do your creative work; have you any idea?'

Well, getting original ideas, that is, useful original ideas—useless original ideas are ten a penny of course—usually follows a certain set of general principles in most fields, including the arts. I first learned about these when we had a seminar in Aspen, Colorado, more than twenty years ago. There were two painters, a poet, a biologist, and one theoretical physicist—me. We were all talking about occasions on which we had gotten useful, creative ideas in our different fields, and it was remarkable that our descriptions were all very nearly isomorphic. In each case there was a difficulty in the way, and in each case the person thought very hard about how to overcome that difficulty. In theoretical science it would often be the difficulty of explaining some experimental data that didn't conform to existing theory. In painting or poetry it could be the difficulty of expressing something by those means that seemed to be available. And in each case the person devoted an enormous amount of thought to overcoming the difficulty but was unable to do so. A point was reached where further thought seemed to be useless. And yet, outside of conscious awareness, the mind seemed somehow to go on working on the problem, because at some point later on—say while running or cycling or shaving or cooking—the person would suddenly get a useful idea about how to overcome the difficulty. I was much impressed with this discovery of a process that was common to such varied fields. What I didn't know was that people had called attention to it a long time before. For example, Helmholtz, the great Prussian army surgeon, physiologist, and physicist of the nineteenth century, had described it, and he had called these three stages saturation (saturating the mind with the problem), incubation (waiting for the mind to incubate a solution), and then illumination (seeing the useful, creative idea suddenly appear at some odd time). The mathematician Henri Poincaré wrote about it later on, adding the fourth stage, which is important, though fairly obvious—verification (checking to see if the idea really works).

'You refer in your writing to one of your discoveries coming from a slip of the tongue.'

Yes, that was the one that I spoke about at the seminar in Aspen in 1969. But you have to remember that Kekulé, the chemist, claimed long after the fact that he had come to the discovery of the form of the benzene molecule through a vision when he was half asleep, which is in some ways even more remarkable. A number of historians of science doubt that his story was correct, but he claimed that he had been trying to explain the form of the benzene molecule using carbon chains, with hydrogens attached, and that while he was doing so, he had fallen into a 'half-sleep' and dreamt that the carbon chains turned into snakes, one of which bit its tail, so that he woke up with a conviction that the benzene molecule was a ring.

My experience, involving the slip of the tongue, went something like this. I was trying to account for the so-called strange particles. These were particles discovered originally in cosmic radiation during the '40s and '50s, and they had the characteristic that they were produced copiously but took a 'long time' to decay, around 10^{-10} seconds. That is considered very long by elementary particle theorists because it's 10^{-14} times the time it takes light to cross a nuclear particle. The explanation, I thought, was that the strange particles were strongly interacting (so that they could be produced in great numbers) but that there was some forbiddenness rule preventing the strong interaction (which includes the force that binds protons and neutrons together in the atomic nucleus) from causing strange particles to decay. They would then have to wait around for the weak interaction (responsible for phenomena such as beta-radioactivity) to make them disintegrate. That would explain the long lifetime. However, when I tried to identify the forbiddenness rule involved with the principle called conservation of isotopic spin (a correct but approximate law for the strong interaction), I ran into trouble because the forbiddenness could be circumvented by electromagnetic effects. In the presence of electromagnetism, which is not so very weak, the forbiddenness for the strong interaction can slow down the decay by only a relatively small factor.

When, in May 1952, I visited the Institute for Advanced Study at Princeton, where I had been working a few months earlier, people asked me to talk about this matter. They hustled me into the seminar room, all the theoretical physicists poured in—including Robert Oppenheimer, the Director of the Institute—and I had to go to the blackboard and explain what my idea was and why it wouldn't work. Now, at a certain point in the talk I had planned to say: 'Suppose this particle has isotopic spin five halves. The strong interaction would not permit it to decay and we would have a valid explanation of its long lifetime—except for the fact that electromagnetism ruins it.' However, instead of saying, 'Suppose this particle has isotopic spin five halves', I said, through a slip of the tongue, 'suppose this particle has isotopic spin one'. And then I stopped because I realized that with isotopic spin one, the objection involving electromagnetism disappeared and the theoretical idea would work.

'And that was the answer?'

Yes, that was the the correct answer.

'It is, to say the least, rather unusual to make a discovery in the middle of a lecture.'

[Laughs]. Of course, at the Institute for Advanced Study the atmosphere was such that people immediately pooh-poohed the idea, and I actually didn't publish it for a year and a half afterwards.

'One of the features of your area of particle physics is the language: strangeness, colour, charm, quarks and so on. Is that due to you?'

Some of it is, yes.

'So why choose this rather zany language?'

Well, since I'm interested in languages I'm perfectly capable of inventing, for each of these things, a pretentious word of Greek origin that would satisfy any pedant. But I noticed that the precedents were not very encouraging. Generally, when people have chosen such a name in the past, they have utilized some property that they thought the object possessed, but which has turned out later not to be an important property, or even, in some cases, a correct one. The word 'meson', for example, comes from the Greek word for middle, as it originally applied to a particle intermediate in mass between the electron and the proton. But we now have dozens of kinds of mesons, and many of them are heavier than the proton. So one might as well invent a zany, relatively meaningless name, and have fun.

'Are you still in awe of particle physics?'

Well, nature as a whole can't help inspiring awe at both ends of the spectrum of complexity. At one end there are the fundamental laws that govern the elementary particles and their interactions—beautiful, simple, universal principles. And at the other end, there is the enormous richness of complex phenomena, especially complex adaptive systems with their evolutionary processes. It's so easy to be awed by what science reveals about nature.

'And you continue to be?'

Oh, very much so.

'You're a very powerful person in American physics.'

No, I hardly think so.

'You don't think with success has come power?'

No, I wouldn't mind having some power, but power has never gravitated toward me for some reason. I think a person has to have a little bit of pomposity in order to attract power, and I've never been able to summon enough of it for the purpose. I guess I'm too flippant. I can't resist making wisecracks, and there are always silly people around who take them seriously.

'What has driven you then?'

I don't really know. I've never taken myself seriously enough to think in those terms. I just bumble along, muddling through.

'Is it wanting to be right?'

Ah, that's always very nice. Understanding things, seeing connections, finding explanations, discovering beautiful, simple principles that work is very, very satisfying. But let's not forget the pleasure of enjoying nature directly—walking along a forest trail and seeing (as well as hearing, feeling, smelling, and tasting) the wonders on all sides, or wandering in some expanse of mountain scenery and looking at the plants and the animals and the views. Perhaps scientific discovery matches that for pleasure, but I'm not sure.

'So, you're a real romantic?'

Well, I'm a mixture. Certainly one strain is very romantic.

SHELDON GLASHOW
*was born in 1932 and is Higgins Professor of Physics at
Harvard University.*

CHAPTER 18

The weirdest of fancies

Sheldon Glashow
Theoretical physicist

WHILE science seeks to find order in nature, the means to that end is usually far from orderly, and in the history of science there can hardly be a greater contrast than between the elegant architecture of modern physics and the piecemeal *ad hoc* manner in which it was constructed.

In 1979 Sheldon Glashow, Abdus Salam and Steven Weinberg received the Nobel prize in physics for work which showed that electromagnetism, one of the four fundamental forces of nature, and the so-called weak force, which acts at the sub-atomic level, were aspects of a single, more basic principle. It's only a matter of time, most physicists believe, before the other two forces, gravity, and the strong nuclear force, are brought into a grand unified theory.

The present picture of the physical world is one dependent on three fundamental constituents. There are quarks, of various sorts, with whimsical sounding names, up, down, strange and charm. They make up the particles of the atomic nucleus, and are governed by the strong force. There are leptons, lighter particles, such as the electron which are involved in weak interactions; and bosons, which are the agents or carriers of the basic forces. They include the photon, and the more prosaically named W^+, W^- and Z^0 particles.

Sheldon Glashow's involvement in the problems of the weak and electro-magnetic forces began by chance, and was abandoned more than once. The whole story is one of unrecognized connections, discarded papers and forgotten theories. I intended that we should try and unravel it, and, to begin at the very beginning, wanted to know what had brought Glashow to physics in the first place.

That's kind of a tough question. I was brought up partly at the beach, and each summer we would watch the tides. I was fascinated by this regular behaviour, I was fascinated by what little I knew about astronomy from looking at the sky. I remember perhaps one of the most significant moments was in Junior High School when we were taught about the revolution of the Earth about the Sun, and the rotation of the Earth about its axis, and I realized a very strange thing. I realized that the moon—the man in the moon—always faces the Earth. It perplexed me because I would have thought that the moon would be spinning about its axis, as it pleased, and as the lunar month progressed we would see different sides of the moon, and yet we could only see the same side of the moon. I asked my teacher, and the teacher, of course, didn't know the answer —I didn't learn the answer for another ten years or so—but she said, 'That's a good question', and she really encouraged me, and made me think that I knew something about science because I was able to ask the right question, and I just, from that point, went on asking questions because I was praised the first time.

'And from then?'

Through High School, where I passed through various stages of biology, and chemistry of various kinds, but I finally realized that physics is the most important science, and got interested in that and started trying to learn quantum mechanics and relativity on my own. While the teachers were burbling about levers and inclined planes, I was thinking, to hell with this, I want to understand particles.

'Were you really a bit of a prodigy, a child prodigy?'

Not really. I didn't particularly graduate in the top of my class at my High School, perhaps the top 10 or 15 per cent, but there were a lot of us who were prematurely interested in science, and knew that we wanted science as a career. We formed a sort of group that reinforced one another. One day somebody would come to me and say, 'Hey, I learned calculus of variations yesterday, it really isn't very hard.' So I would go home and study for a week, and say, 'oh well, I learnt quantum mechanics, so there!' We didn't really learn either of them, but we had some inclination as to what it was about.

'How old at this stage?'

Sixteen, seventeen. So we tried to learn as much as we could in the absence of teachers, and in the absence of knowing what the right books were, but we did as best we could, and we went to college which, of course, straightened us out.

'But then what made you choose theoretical physics, rather than experimental?'

I was never into doing careful experimental things. Some people get into physics by building telescopes and getting very excited about astronomy. Others in my time would build radios and other electronic devices of various kinds, but every time I tried building something it wouldn't work, and if I took parts out of one clock, for example, I'd never get it back together again. So I just knew it wasn't for me. And college reinforced that. We had a laboratory course in first-year or second-year physics, and they asked me to leave. I insisted on dropping lead bricks on top of Geiger tubes and so on.

'So, it really wasn't a choice, you didn't have the option of becoming an experimentalist?'

I'm not really as good with my hands as my father was, for example.

'Why, what did your father do?'

He was a plumber.

'Did you go from the very beginning for the big problem, or what you thought to be the big problem?'

Well, at the very beginning I had to go to college, where I had to learn a lot of things that I didn't really know, even though I might have thought I knew them. I had to go to graduate school to find out what the problems are. You can't, as smart as you might be, you can't just jump into something like theoretical physics until you know what's been done before.

'But what did you jump into with your first research problem?'

Well, it's very characteristic of theoretical physicists that they continue to work along the directions of their thesis, and that happened to me. I had the good fortune to study under Julian Schwinger, who is a very famous man.

'Where was that?'

At Harvard, as a graduate student; and after I was there about a year or so, I, together with about ten of my colleagues, went to Schwinger and said we wanted to be his students. He took remarkable numbers of students at the same time. So he lined us up against a wall, and this incredibly brilliant man started giving problems, sequentially to ten different people. I was toward the end of the line, and there was something he was peripherally interested in, which was the parallels between the weak force and the electromagnetic force, and he suggested that I worked along those lines, without giving me any specific ideas because he really didn't have any at that time. So that's what I did. For a couple of years I thought about how you could create a unified theory of weak and electromagnetic interactions.

'So, from the very beginning you went for that problem?'

That was really the substance of it, yes.

'Was it luck?'

It was lucky, it was good. There was a real problem there.

'But now, how did you work? Did you work on your own, in a group, did you go home and just think? How does one do that, how does one approach that sort of problem?'

Well, one participates in the intellectual affairs of the place you're at, which in this case was Harvard, and goes to seminars and talks with friends. But mostly, on your own problem, you think about it and you work and you scribble, and you try to get somewhere. I didn't get very far, but I accumulated enough material to form a passing fair thesis and left. I was intending to continue work along this line and other lines in Russia. So, I went to Copenhagen to wait for the visa. I had written to Bohr saying I'd like to get to Russia, but as Copenhagen is most of the way there, I would wait there until I got the Soviet visa. Needless to say, the Soviet visa never came, and so I spent, in the end, two years in Copenhagen, following up on the thoughts that I had as a thesis.

'Did you make real progress when you were there?'

Yes, although I didn't know it at the time, and nor did anyone else.

'That's a puzzle, I don't quite understand that.'

At that time it was just a dream that you could create a theory which would unify weak and electromagnetic interactions. There were a number of steps missing. The work done by Steve Weinberg, the idea of quarks and lots of other things, had to come; spontaneous symmetry breaking, Geoffrey Goldstone's work; but nevertheless you could say things about such a theory if such a theory were ever invented. So you could say, for example, that there would be a W boson, W^+ and W^-, that would go with the photon, and then the question was, 'is that all the bosons that you would have to inject into the theory?' And the answer was, 'No'. What I realized is that there had to be this neutral heavy particle, Z^0 it's called today, as well, and that the mass of the problem insisted that there would be a Z^0 as well as a W. The group structure of the theory, so to speak, I established back then, in 1961. But there was no theory. No one believed in a unified theory of weak interactions; no one believed in W bosons, let alone Z bosons; and lots of things had to happen after the early '60s when I did this.

'OK, so then how did it go from there to the unification?'

Well, let's see if I can reconstruct that. I think I, together with the rest of the world, have mostly forgotten entirely about this particular game that I played.

'And went on into something else?'

Went on into other things. I made my name in other little bits and pieces of physics which, by 1966, were good enough to get me tenure at Harvard.

'But not the work that eventually got you the Nobel prize?'

Not at all, no mention of this work in 1960 which was to be important later on, no mention. Let's go back to '61 when I spent some months at CERN and Geoffrey Goldstone, the famous British physicist who's now, of course, right down the river at MIT, had invented spontaneous symmetry breaking. He and Salam and Steve Weinberg were very interested in this subject, and we didn't realize, nobody realized, that it was going to be very, very important for the theory of weak interactions. I was with Goldstone, looking over his shoulder when he invented spontaneous symmetry breaking, and we never put things together. Then I went to a summer school in Scotland, in fact; the first 1961 summer school in Scotland . . . it gets very complicated . . . I'm trying to reconstruct all these strange things I did, trying to get to the point of 1964 when Peter Higgs invented the Higgs' Mechanism. I was there at the time too, and just after he had made this great invention. Peter Higgs, a Scotsman, takes the work of Goldstone, the Englishman, and shows how, if you combine together two exceedingly abstract and crazy bits of physics, something else even more crazy comes out.

What comes out, we know today, is that the W boson gets its mass. But he didn't realize this, and we were too busy chasing the same girl to talk about physics. So again, if in 1965 or so, we had only chatted about what his invention could have done, we would have created the theory in full measure at that time. But no, he did his thing, and he forgot about his work. Then, Steve Weinberg and Abdus Salam, in '67, quite independently, both realized that this mechanism of Higgs, which makes use of the ideas of Goldstone, and also folds together with work that was done at various times by Weinberg and Salam, could all be used to create a sensible, or what they speculated to be a sensible, theory of the weak force for leptons. That is, acting on electrons and neutrinos, not acting on nuclear particles. It was a little part of a theory, what we would today call a toy theory. So, Steve published a little article called, I think, 'A Theory of Leptons'. Now, that went over like a lead balloon too. I mean, Steve forgot about that work, the world never read it, and time went on. Time went on until 1971.

'Did you recognize its importance?'

Absolutely not, I didn't even read the paper. So, it was very strange. Now wait, in 1970 a Greek physicist named Johnny Leopolis, who has become French, and a seemingly Italian physicist named Luciano Miote, who was in reality San Marinan, and thus exempt from the Italian draft, were visiting Harvard, and together we did something interesting having to do with charm and the fourth quark, a subject that I'd been interested in before. In the process we realized that that would be very good for my old theory of weak and electromagnetic interactions, and we went to MIT and talked to Steve Weinberg about how we could extend the unified theory. He didn't seem to care, I mean, he had forgotten his own work, or decided it was wrong and irrelevant. We wrote it up and referred to his paper. We didn't read it.

'But you went to talk to him?'

Yes.

'And he wasn't at all interested?'

No, not at all. No one was interested until 1971, when a Dutchman, Gerard t'Hooft, a young graduate student at the time, independently did my work of 1960 and independently did Steve's work and Salam's work of 1967. He recreated this theory, but more importantly, with the assistance of T. D. Veltman, also in Holland, proved that the theory made sense, made mathematical sense. Then there was an explosive interest in what was then called the Weinberg–Salam Theory. It predicted neutral currents, and the neutral currents were found at

CERN in 1973, and confirmed in the States in the same year. The search for the W boson and the Z boson became important, everyone became deadly serious about this grand new theory. It was funny because all of the preceding steps—the algebraic structure of the theory that I did back in '60, the nature of spontaneous symmetry breaking, and how it gives mass to the Ws and Zs (in 1967), the fact that you could extend it to a theory of protons and neutrons, making use of charm and quarks—all of these bits and pieces fell together in November of 1974 with the discovery of the J psi particle.

'I must interrupt. Did you understand now that it was important?'

We're now in November of 1974.

'I want to know, had you clearly recognized how important your early work was in relation to the theory?'

Yes, but the key to it in my mind was that you had to believe in quarks. It had to be more than a toy theory of electrons and neutrinos, it had to deal with neutrons and protons as well, and back in 1970, with my Italian and French/Greek colleagues, we had decided there had to be a fourth quark, and there was no evidence for this fourth quark. I had gone to a meeting in April of 1974, in Boston, saying that this particle had to be discovered and it should be the main thing to be searched for by experimental physicists throughout the world, and I was nagging them to find me this particle. And then, in November of 1974, uninfluenced by me—it was different people—it was found. It was found as the J particle at Brookhaven National Laboratory, and as the psi particle, independently, and on the same day, in California.

'On the same day?'

On the same day. Here's how I found out about it. I was at home, the way I was this morning, and I was suddenly called up, would I come to a meeting of a small number of physicists at MIT immediately. So I went to MIT and they told me about the discovery of this new particle; and it was fabulous, it was an exciting new particle, it was not like the 100 particles that had come before. It clearly meant something, but it wasn't entirely clear just what it meant. So I got back into my car, figuring I would bring the news of this particle that I had just been flattered to be picked out to learn about, to my colleagues at Harvard. So I came into Harvard where everybody was jumping up and down because they had just heard of the discovery of the California version of this particle. And we talked about it, everybody was screaming and shouting and jumping up and down, and within a few hours my friends and I had convinced ourselves that this was it, that this particle was clearly a particle made of charmed quarks, made of a charmed quark and its anti-quark, and its properties told us that all of these crazy ideas were making sense.

'You keep talking about crazy ideas.'

Well, first of all, the quark was a crazy idea. There were these particles that no one had ever seen, and according to the theory today, no one ever will; that carry fractional electric charges, charges smaller than the charge of the electron, and that they were the constituents of all of these hundreds of particles that had been discovered. They came in three varieties, called up, down and strange, according to Murray Gell-Mann, and I was an advocate for many years that there was a fourth family, a fourth quark called the charmed quark. Now, what held them together? Well, first of all, although Gell-Mann wanted three of them to stick together to form a proton, for example, they had to behave kind of weirdly to do this. It seemed that they didn't satisfy the laws of quantum mechanics and relativity, so that meant there was something fishy. Some people said the quarks had colour, that not only were there up quarks, down quarks and strange quarks, but each one came in three indistinguishable varieties, that for some unknown reason were called the three colours, three primary colours, and that they could only stick together to form colourless particles.

'It's in that sense that you mean the theories were crazy.'

Yes; all kinds of things like that were present in the theory, highly speculative things. All right, I mean, there are these colours. If you would simply take the kookiest ideas of the early 1970s and put them together, you would have made for yourself the theory which is, in fact, the correct theory of nature. So it was, it was like madness, it was everybody's weirdest fancy was right.

'And so from '74 to really getting it all together, one year?'

By '75 it was clear that we had a theory in hand that would do all kinds of things. I remember what convinced me was that we had a bunch of particles, one of which was called the sigma, another was called the lambda particle, and I said to my colleague, Howard Georgai, that I would believe all this crazy stuff if he could convince me that this theory which, in principle, would answer questions like, 'Does it explain why the one particle is heavier than the other, and how much?' And he went home and he showed that, indeed, that was a consequence of the theory. We were then convinced that there had to exist dozens of new particles that hadn't been seen, called charmed particles, and we could predict, and other people could predict, what their properties would be.

So, it was a heroic time when everybody was talking about the properties of particles that we knew existed, but the rest of the world was not yet so convinced, and they had to go and find it. There was a period of eighteen months or so when we knew the truth at Harvard and at a few other places in the world, but not, for example, at Stanford. The Stanford people were so confused that they put out a very long paper, a collaborative paper of the theorists, with perforated pages, with the notion that as soon as some idea had been proven wrong you could tear it out and throw it away. Eventually, of course, all that was left was the binding and the sound of laughter from Harvard, because we knew how things were going.

'Looking back, do you think that you missed out on something by not realizing at that early stage how important your ideas were? I know hindsight gives one wonderful wisdom.'

We all missed out. I mean, everyone is angry with himself at all points. My French/Greek friend could have put everything together years and years before if he had simply known about the various developments that were out there in the literature. Peter Higgs could have done more than he actually did, although he did quite a lot.

'Could you yourself have done more?'

I certainly think so, but we didn't.

'To what extent were you put off by experiments? I mean, there must have been some experiments that didn't fit the theory.'

Yes, but we had to be very firm.

'What do you mean by "be very firm"?'

Well, you recall that I went to a meeting in 1974, and predicted that charmed particles would show up. In fact, I said if they did not show up by the next meeting in this particular series, I would eat my hat, but that they would have to eat their hats if it did. Then in 1977, there was a sequel to this meeting in 1974 of meson spectroscopists—these people who were looking for particles in the wrong places because they weren't convinced that these particles existed—and they conceded that these particles had been found, although not by them, and they ate their hats. The organizer of the conference provided foul-tasting candy hats which he distributed to all of the members of the audience, and I watched as they ate them.

'Terrific. So you really never had any doubts that your theory was right?'

Not of that particle at that time. At earlier times it was not clear, and even at later times there were experiments that didn't quite work. In 1978 there were a few wrong experiments that had been done which didn't fit with the theory, and it was very upsetting to all of us. Some of us tried to make baroque modifications of the theory that could fit the data—I must say I did that too—but eventually the experimenters found their errors, and everything has worked perfectly ever since. And we are stuck, I use the word advisedly, with a theory, or a model, something we call the standard model of elementary particle physics, which so far as we can tell, and with the expenditure of enormous amounts of funds, works perfectly. We haven't found any flaw.

'Just one thing. One often thinks of physics, or physicists, theoretical physicists, being driven by beauty and simplicity. The way you've described it, it's driven by a curious sort of madness.'

Yes, it is driven by madness. But the thing that came out is quite beautiful when you finally get at it. Suppose you turn around and teach physics as it should be taught; suppose, instead of historically, you simply start at the beginning and say, 'This, like it or not, is the way things seem to be', the theory from that point of view is very straightforward and seemingly logical.

'But you weren't driven in getting to the theory by ideas of simplicity and beauty, or were you?'

This is not a theory that you could, that Einstein could, possibly have invented. That is to say, it does not come from pure mind.

'It comes from what sort of mind, then?'

It comes from experiment. It comes from what we learn about nature by doing experiments.

NICOLE LE DOUARIN
was born in 1930 and is Professor at the Collège de France and Director of the Institute of Cell and Molecular Embryology at Nogent, France.

CHAPTER 19

Directly to the heart

∽∾

Nicole Le Douarin
Developmental biologist

LOUIS Pasteur's dictum that fortune favours the prepared mind is almost a scientific cliché. It is nevertheless undeniable that the events leading up to important discoveries often seem to contain an element of chance, but equally true that chance is transformed into luck largely by the insight of the recipient.

Nicole Le Douarin's early success seems to be no exception to this rule. She was, and still is, concerned with the problems of development. How, from a single cell, the fertilized egg, billions of different cells arise and organize themselves into heads, eyes, arms and legs, and so on. How, for example, early in development, cells belonging to the neural crest in the region corresponding to the back of the head move forwards to form much of the face. Her first major contribution was to realize that there were certain differences between chick and quail cells which could be exploited to track such cell migrations. She grafted the corresponding quail cells into chick embryos, and showed that as the tissues developed, the grafted quail cells acted as visible labels, or markers, whose destiny could easily be followed. This work led to a transformation in our understanding of cell migrations, particularly those involving the head and nervous system.

Nicole Le Douarin is now one of the only two women professors at the august Collège de France, and has become director of the same Embryology Institute in Paris where she trained under the embryologist Etienne Wolff. She is a foreign member of both the Royal Society and the United States National Academy of Sciences, and was recently awarded the prestigious Louis Jeantet prize. A glittering career indeed, but was any part of it really based on luck in any important sense? Was she prepared, for example, by a family background in science?

Not at all. My mother was a teacher and my father was in business, so there was no science or research at all in my background.

'How, then, did you get into science?'

I came to science after being a teacher myself for several years. I was a High School teacher for eight years, and it was only six years after I started that I decided that I wanted to go back to university, and that I was not at all

satisfied by a career of teaching. I decided to try and find a laboratory to prepare
a Ph.D.

'Starting research after six years' teaching is really quite a long delay. Most people
who go into research really start rather young.'

Yes, but I was very busy at that time. It was the time when I had my two children,
and also, initially, I was very interested in the process of teaching. I had to learn
how to teach and that was interesting. Even so, after a while I realized that this
was something which was not what I wanted to do for my entire life. I wanted
something where I could be more dynamic and more inventive.

'So, what did you choose to work on?'

It was rather difficult to find a laboratory to work in, since, being a teacher, I
had to give lessons during a large part of the week. It was really hard for me to
find a suitable laboratory. But I had a friend who was with me at the university,
and I told her my problem. She said, 'I'm going to talk to my boss, Professor
Etienne Wolff, who is very nice and does interesting work. Perhaps he will be
interested in your coming.' This led to an interview with Wolff. I told him,
I am not asking you for a position, because I have one already. I am asking
you for a project and some place for me to work. I will come as often as I can
when I have time off from my teaching duties. He accepted the idea and I was
very happy.

'Was it just this contact with your friend that brought you to his laboratory?'

No. I explored other possibilities, but when I met Etienne Wolff, and found out
what was going on in his laboratory, I realized that development was really a
subject that appealed to me.

'What was it about development that appealed to you?'

It is fascinating to start with an egg and twenty-one days later you have a chick.
You watch that happening in the egg and you realize, even if you are very
ignorant, that the process is very mysterious. This is what fascinates me, and still
does. We are still very far from understanding what is going on. Development
of the embryo is one of the most challenging and fascinating problems in all of
biology.

'What did you begin to work on?'

The project that Etienne Wolff gave me was to study the development of the
digestive system of the chick.

'And did you find what you had been missing in teaching?'

Yes, absolutely. I realized immediately that this was what I wanted to do. What
was particularly important is that you were required to have a lot of initiative.
If you don't have initiative you are not a research worker and it's better to do
something else. You are required to imagine what you are going to do. You

are free and have complete liberty in thinking about how to proceed. This is something fantastic, and is what I was looking for. Probably all the people in research think this way.

'When did you come to make your big discovery of the quail cell marker?'

I made the discovery one year after I got my Ph.D. degree. I was then an associate professor in the University of Nantes. In fact, there was a lot of serendipity in this discovery. The crucial event was that I was given quail eggs to work with. Quail at that time was not used in experimental embryology. But it so happened that a geneticist was working with quails. A female quail lays one egg per day, and since he had a very large colony of quails he had an enormous amount of eggs. He didn't know what to do with all these eggs, so he suggested that quails' eggs be used in embryology laboratories in which the chick egg was the normal experimental material. And, indeed, many people at that time started to work with quails. But I must say that no one else did notice the peculiarity of the nucleus of quail cells, that made them clearly distinguishable from chick.

'Did you realize at once how important this was?'

I think I did. When I worked on the development of the digestive system I became very interested in movements of cells during morphogenesis. I tried tracing the movements of embryonic cells with various means which were very imperfect at that time. I had used vital dyes, or carbon particles to mark the cells. But when I saw the quail nucleus I immediately had the idea that if I combined chick and quail tissue I would be able to follow the movement of the cells. I remember when I did the first experiment. I grafted the neural tube of a quail into a chicken, we could see the migration of neural crest cells all over in the chick embryo. I asked the whole laboratory to come and see the cells that had migrated because it was so striking. There is real beauty in these preparations because of the clarity of the difference between these two cell types—chick and quail—for an embryologist it is something that really goes directly to the heart.

'It must have been very exciting.'

Extremely exciting, and not only for a short length of time. The excitement lasted for a long time and it even continues today. I have communicated this excitement to many people.

'In what sense do you see your discovery as involving serendipity? It wasn't as if you weren't interested in the problem in general. I'm somewhat against the idea of serendipity since it implies that you discover something that you weren't really looking for. But you really were looking for something which would enable you to trace cell movements.'

I am not an English speaking person, perhaps I don't use the word serendipity properly. What I mean is that there was some luck in the beginning because I was given these quail eggs. Of course, many people had the same opportunity and did not see anything. I was prepared to look at the nucleus, and I had the idea of using it because I was interested in cell migrations already. So I agree with you, it is not serendipity in this sense.

'And it's not even really luck. It always seems to me it's always the good scientists who are the luckiest.'

[Laughs]

'Do you think there is much luck in science?'

Not much. A lot of work is needed, much more than luck certainly.

'But what do you think that your particular skill is? Are you particularly skilful at operating on embryos?'

Well, yes, I like that very much.

'Do you still do it?'

Yes. Not as much as I would like. But I hope that when I have less responsibilities—I mean responsibilities that I cannot escape from at the moment, but hope to be able to escape from later—I will go back to the bench more. I really like that, it is very amazing for me to do that. When I do it I have the opportunity of having new ideas and formulating other problems. It is very inspiring for me to do the experiments actually.

'You spoke about the aesthetic delight when you saw those sections with your quails cells. Is that part of the pleasure that you get?'

Yes, absolutely, certainly. This is something that is always renewed because when you do an experiment, and when you find a result that you expected, or you find something that you did not expect, you have the impression that you are approaching something real, approaching the realities that you are seeking. It is something which is unique.

'But what about the aesthetic pleasure? Do you actually think embryos are beautiful?'

Embryos are beautiful and if you have the chance to look down the microscope you have a fantastic pleasure which is renewed each time. This is especially so now because we have new techniques to look, not at cells as we did twenty years ago, but to look at molecules. We look at receptors, we look at gene activity.

'Do you think it's as beautiful as the old, or the more classical preparations?'

I think the beauty is not only in the colours, the beauty is what is intellectually behind what you see.

'You've got a group of about fifteen people. What's your relationship with them?'

That's interesting, because the group is composed of people of different ages.

There are some people who have been with me from the very beginning, and with whom the interaction is really a delight because if it had not been so they would not be with me. So there is really a constant exchange of ideas, a lot of mutual confidence, and this is the backbone of the group. And then there are young people who come from the university who are very fresh, very enthusiastic, who want to learn, who want to do things with the latest techniques which appeared in the market. This is extremely important for the whole group since the vitality of the group comes from these young people. So, all together it is one of the most pleasant things in my work, this interaction with the people in the lab. It is something which is completely informal, we are all together, we talk to each other all the time, and this is what I really enjoy the most.

'I think that's a very big difference between the arts and the sciences. For example, when somebody does a Ph.D. in the arts, they do it in a very isolated manner, and I think people don't fully realize that when one does, for example, a Ph.D. in our subject, that they're working within a group. It's a much more social activity.'

That's absolutely right, yes. That is something very striking. It has advantages because of the interaction between people. It has also some disadvantages, I must say, because I realize that each person who has participated in a work, and who at the end signs the paper, does not have the feeling of being completely responsible for the results, as has the person who has written a book all by himself. But this is how modern biology works.

'Am I right in thinking that most of the papers from your group, then, are multi-author?'

Yes, but it was not like that when I started in research. The first papers that I wrote were all by myself. But now it is absolutely necessary to have some sort of multi-disciplinary approach to nearly every problem.

'To what extent are they free in your group to design their own experiments?'

Well, they are very free. It's extremely dangerous to tell students, 'Don't do that', because one of the things with biology is that you never know what is going to happen. It is so complicated, there are so many parameters that you ignore in life science. I think it is wise to have this attitude.

'Do you think the young really do have that advantage of coming in uncontaminated with the past, as it were?'

Absolutely, absolutely. We contaminate them with the past when they arrive, but this contamination has something good, certainly, because, of course, science is history, but I think that this is a big advantage, for them and for us.

'You travel a great deal. Do you enjoy that?'

Yes, yes I enjoy it to a certain extent. This is a difficult problem. It is very important to travel and to be aware of what is going on in science, and also to let the other people know what you are doing. It's clear, that science progresses so fast now because of the increasing communications between people who are involved in science. But I enjoy travelling only to a certain extent. When you travel too much it is disturbing for the work, disturbing for your personal life; it is tiring also, so you have to find a balance.

'So, because you're familiar with science in other countries, do you think that French science is more conservative, or different in some other way from science in other countries?'

I think that French science has evolved a lot during this last twenty years, and one thing which has made the science evolve significantly is increasing communication. First of all because people travel, and essentially because young people, when they are at a formative stage, go abroad. The postdoctoral system, exchange of people during the postdoctoral period, is something very important.

'You're really saying that French science has become internationalized?'

In our field, at least, I have witnessed that during this last twenty-five years or so.

'But, of course, it also means that perhaps the French language has lost out a bit?'

Absolutely.

'Because there was a time, I remember, when the French tried to resist this and even published in French in international journals.'

No, that does not happen any more. But I am sure, I am convinced, that in the French population there is a large part that does not really agree with that, does not accept it, and regrets it.

'You are talking about the French scientific community?'

No, the French population in general. I am sure that there is a regret that French is no longer a language that many people out of our country can understand.

'Then what about the French public's attitude to science in general?'

Oh, they are very much interested in science, and they understand that research is something important for the society, for the welfare of humanity. I think they are very much in favour. The increase, for example, in the budget for research is something that has occurred during these last years significantly.

'You're almost saying that there isn't that sharp division which some people see in the United Kingdom, for example, between the humanities, on the one hand, and science, on the other, in France?'

I think that people in France, the general public, cultured people, read a lot about science. There are good journals which are popularizing our science, and there are also some very good books.

'To what extent do they really want French science to succeed, to be proud of French science?'

They are very proud. When there is a Nobel prize in France people are absolutely happy, absolutely.

'Now, Nicole, you've been enormously successful. Recently you got the Louis Jeantet prize. You were elected a foreign Fellow of the Royal Society, and you were also elected a foreign member of the U.S. National Academy of Sciences. First of all, I'm sure you were very pleased about that, but how important are these things in a scientist's life?'

It is very nice, but it does not really change the daily life, at least for me. You know very well, because you are a scientist, and a very successful one too, that people in science are anxious, and especially when you are leading a group, you are anxious about whether you are going to solve a certain number of problems; whether you are working in the right way; whether you are doing well; and whether your projects are good for the future. I must say that the fact of having all these really great honours, like to be elected at the Royal Society, was really a fantastic surprise, and a great pleasure. But it does not change this anxiety, at least for me. I am as anxious now as I was when I started.

'Do you really mean that?'

I really mean that, and I think it's terrible. I would like it would not be like that, but it is like that. I think that if you are not anxious you are not a researcher, you are not a scientist. Because if you are very quiet and very satisfied then why try to find what is not known? So if you really want to understand more it's because you are fundamentally anxious, and the fact that you want to do that makes you even more anxious.

'But what is the nature of the anxiety, that you're going to fail?'

No, but that you want to know what is going to happen; you absolutely want to find the best way to do it; not because you are going to fail. I think that this is the same anxiety as the tennis player has before a contest, he wonders whether he is going to play well.

'And the prizes really don't remove this entirely?'

Maybe I do not get enough of them [laughs]. No, what I mean is that I think that this kind of anxiety is extremely important for the continuation of the work. I think when it is completely removed it's because you are ready to retire. Most of the scientists who are really researching in their soul never retire.

REFLECTIONS

GERALD HOLTON
was born in 1922 and is Professor of Physics and of
History of Science at the Jefferson Physical Laboratory,
Harvard University.

CHAPTER 20

Other minds

❧

Gerald Holton
Physicist and historian

HOWEVER much energy they might devote to it, historians and psychologists will never fully be able to explain why it was Einstein who came up with a general theory of relativity, any more than they will be able to say why Tolstoy wrote *War and Peace*. Even so, attempting to cast light on the social and personal factors involved in the process of creation is a reasonable endeavour, and has proved, in the arts at least, to be a rich academic hunting ground.

Scientists, on the other hand, have tended to be suspicious of such activities. There is, perhaps, a feeling that unless someone has actually done serious scientific research, he or she cannot possibly understand it. That's why Gerald Holton's work is so interesting. He is both a successful physicist and a historian of science. His studies of Einstein's correspondence, and the laboratory notebooks of Robert Millikan, the American physicist whose famous oil-drop experiment determined the charge of the electron, have provided convincing insights into the way great science operates. His work on other founders of modern physics, such as Rutherford, Planck and the Italian nuclear physicist Enrico Fermi has also, among many other things, revealed much about the phenomenon of great scientific schools.

To begin, I asked him whether doing science and doing history had much in common?

In some ways not at all; but in another way, they are the same: in both cases you simply try to give your obsession free rein and try to solve puzzles, and in that sense the human emotions that are involved are much the same.

'But the methodology?'

The methodology is, of course, completely different. In physical science you have an external chastising arbiter. You are found out more rapidly if it's wrong, if you're in science; in history it takes longer.

'What did you actually set out to try and find out, then, when you became a historian?'

As so often, it has much to do with accidents. I think the key moment came on the death of Einstein in 1955. His biographer, Philip Frank, who was also a professor and colleague at Harvard, suggested that together we should bring out some review of Einstein's work. This was the first time that I took history seriously; seriously enough to go down to Princeton to look into Einstein's correspondence which was assembled there. His scientific correspondence was immense, and it opened my eyes, in a way that I hadn't realized before, to the amazing fruitfulness of trying to do history of science through the letters, the manuscripts, rather than simply through their publications.

'What did you try and find out from that correspondence?'

I wanted to know, ultimately, what was his research programme. He was, of course, amazingly productive; there are almost 300 papers, and in many fields, and I wanted to know was whether there was something in common among them. Was there one great research project going on?

'And what was the answer?'

The answer was, yes. We now know that Einstein was, in fact, obsessed with relatively few questions throughout his whole life. One of them was whether there had to be a choice between the different world views with which the nineteenth century was preoccupied, or whether one could find one that makes a bridge between all of those claimants. That was, in a sense, the ambition he once confessed to his research assistant Ernst Strauss, when he said, 'What really interests me is whether God had any choice in the creation of the world'. Just how many different things had one to invent, so to speak, in order to have a physical universe. I think his driving motivation was very close to that. Is there a unity underneath all the phenomena?

And then in addition to it, and almost following from it was the question: to what degree can a few simple thematic notions explain a great deal of physics? The notions of the continuum, of causality, of symmetry, things of that sort, perhaps a dozen of such fundamental ideas, can one use them in all parts of physics? And he showed that one could.

'Was Einstein unusual in having such ideas, or having such a research programme?'

I think he was unusual in the intensity and longevity of this research programme. It comes through from the time when he writes his very first paper telling his colleagues that he is trying to find a way to think through the problem of the forces responsible for capillarity by making an analogy with Newtonian gravitation. He already was unifying different fields, the sub-microscopic and the macroscopic.

'This was when he was still very young?'

This was when he was about twenty-one.

'Einstein seems to have come from nowhere; in other words, he wasn't trained in some great school of physics.'

Indeed, he had a rather limited physics training. He was trained to be a

High School teacher. But for that very reason was much more able to take an independent view of physics. As he once said about Michael Faraday, he would have amounted to nothing if he had gone to college.

'So, then, how did he manage to achieve this?'

Well, that question was put to him: why was it he who made the fundamental breakthrough in relativity theory? And he said, 'that was because I continued to ask the kinds of questions that children are taught after a while not to ask, questions about the nature of space and time'. And I think that that is exactly right; that he preserved the innocence, the ability to look with fresh, you might say almost childlike, eyes, at the phenomena, without allegiance to existing established notions.

'So, when he then sent his first papers in, were people amazed that this nobody, or this person almost from outer space, was having these amazing ideas?'

Until 1905 his papers were fairly pedestrian. He was still trying out his ideas. But in 1905 he sent in three papers, at roughly six-week intervals, to the editor of the *Annalen der Physik*, and as luck would have it, Max Planck was helping the editor in making the decisions. When Planck got the relativity paper, which was completely different in its approach from anything else which was being done at the time, he was so astounded that, so the story goes, he asked his assistant, Max Laue, to take a train to Berne to visit this unknown person to see whether he really existed.

'And did he actually undertake that?'

He actually did. The story is that Laue went into the patent office, from which that paper had come, and asked a young man there, 'Can you lead me to Mr Einstein?' The young man said, 'I certainly can, it is me.' They had a very good chat. Afterwards Einstein said, 'That's the first time I saw a real physicist.'

'That is an absolutely astonishing story. He was totally self-motivated?'

He was very largely self-motivated. Now that we have the letters between himself and his wife and also from the period before they got married, it seems that he shared his early ideas with her, and with only very few other people, such as his friend Michel Besso. He used them as sounding boards.

'With your study of Einstein, and other physicists too, can you give us any insight into the nature of genius, because on all criteria, I assume you would agree that Einstein was a genius?'

That's a very difficult question. I think one must be humble about it and venture a few remarks at most. For example, genius is always measuring itself from the top down, from the prospect of some very large achievement. Genius in physics measures itself with respect not to what has already been accomplished, but, more like Newton, in terms of trying to reach an almost supernatural state of knowledge—supernatural for ordinary people like us—and of being driven toward this almost unattainable goal. Most other people measure themselves from the bottom up, how far they have already come, and are perhaps very satisfied with it: that's deadly. So that's one drive; a mainspring.

Another one is that possessors of genius tend to not see barriers which are obvious to others. For example, think of Freud. The difference between a mature man or woman, and the child; or between daily life and the dream world, all these are obvious to us. To Freud they are not at all obvious; he wants to find a continuity between what to us are completely separate things. The same with Darwin seeing the link between the human species and other species. The same with Einstein joining space and time, mass and energy and so on. So these 'geniuses' are barrier breakers, and that is one reason why we often find it so difficult to follow them quickly. It takes some time to live in a world in which familiar barriers have been broken.

'There's an anecdote about Einstein, it may even come from your own writings, that tells about when Eddington sent the telegram to Einstein saying that he had confirmed the general theory of relativity by showing that the sun's rays are bent during an eclipse, as Einstein had predicted.'

Einstein said, 'But I *knew* that the theory is correct'. This story was told to me by the student who was with Einstein at the time, Ilse Rosenthal-Schneider. She was quite shocked. This was perhaps not a very scientific response by her standards, and so she said to him, 'What would have happened if they had found the opposite?', no confirmation of Einstein's prediction. 'Then', Einstein replied, 'I would have been sorry for the dear Lord—the theory is correct.' That kind of certainty can, of course, be pathological; that's why not everybody who is that certain about his or her ideas is a genius.

'But do you think that is a feature of very good scientists? To really ignore some of the data?'

Even for an average scientist, in practice the principle is *not* that you must constantly find ways to falsify, that is, mainly to find fault with your own ideas. That task is to a larger degree a burden on the scientific community, when the members have to try to test whether something published is correct. While one is working on one's own, often one may have to go through a period of suspension of disbelief in the contrary evidence, the evidence contrary to your hypothesis. That is an often forgotten lesson. So when a student comes to say, 'I really believe this to be true, but the meter readings don't confirm it, what's wrong with me?' You may say, 'Well, let's see first what's wrong with the meter readings. It may be that you don't have a very good ground in your electronics.' In other words, the 'data' do not tell you the truth all by themselves. They may be misleading for a while in an early stage, or in a frontier field. The complementary role of scepticism is, of course, also essential, and, in the end, rigorous proof must be provided. But early on, one has to have some self-confidence and some talent to smell out where it will all lead, before it becomes simple to show it.

'But you're really saying that Popper's falsification idea is an abstract concept that really doesn't work in everyday science?'

It does not work in everyday science as a mandate upon the scientist himself or herself, especially in the early phase of work. But it is perfectly usual for the community as a whole to take up that task.

'Now, you have studied Millikan and his notebooks. Do you think that he took this particular aspect, ignoring some of the data, to excess?'

Well, not if you listen to the sound judgement of the Nobel Prize Committee. The story is this: Millikan had become what one might now call an elderly physicist—he was over forty and had not yet published anything really important. But he had some very good ideas, and he zeroed in on a sound question to ask in 1910, namely, what is the charge on the electron, to a high degree of accuracy, assuming the electron to exist? It wasn't so evident then as it is now. Many of us have done the experiment in Secondary School—it is relatively easy to do now, once you know what has to happen. He invented the method and was taking data for two years. I found his notebooks for a period of one and a half years, and there it showed that out of 140 runs, only 58 really coincided with his presupposition that discrete electrons exist, that they all had the same charge, and that this charge can be found to 0.1 per cent accuracy. That value could not be improved upon for two decades by anyone. However, if he had used those data which he neglected, he would have been quite free to conclude that there is no such thing as an electron, and that the charge can wander all over the place from one experiment to the next. But he knew very well that this cannot happen.

'What do you mean, he knew very well?'

He knew very well in his bones that it cannot happen, in good part because he had, since early youth, been convinced that electricity comes in atoms of charge. He had learned that from one of his great heroes—very appropriately, Benjamin Franklin—who, in 1750, had published the idea that electricity is granular, it comes in particles. That was a thematic presupposition for Millikan. The question then was, could he decently use his data to show whether his presupposition was plausible—though he did it in a manner which nowadays would be strongly frowned upon.

'You mean he just ignored the bad experiments and never published those data?'

Not only that. In his view, as he said in his publications, he had only made runs on the 58; the others just were, so to speak, unfortunate or incomplete incidents in the laboratory that need not be attended to. I think that he really did not believe that he was selecting. He was doing something a bit analogous to what physicists now have to do sometimes, namely, to zero in on the golden events, those which for some reason you believe—and the reasons must be sound and vocalizable—that these are worth putting your bets on, because the other events may have been taken under adverse conditions such as changing voltages, dropping temperatures, things which may all be accounted for in data analysis if one only had enough time. It's a very tricky business to negotiate here between doing very poor science by just finding what you like to find, and doing superb science by having some reasonable intuition what portion of your equipment is sound and when it is sound.

'Do you think that these thematic assumptions, which obviously great scientists clearly adopt, do you think that somehow weakens the scientific enterprise, somehow makes it less reliable?'

When I first stumbled on the large role that these thematic presuppositions can play, it seemed contrary to everything that I had been taught. A scientist is taught to stay away from presuppositions; it's important to keep one's mind neutral, in order to let nature guide you by the hand. But that's not the way things work. Einstein himself said very clearly, and repeatedly, that you have to make presuppositions. And, of course, they can be completely wrong. In Millikan's case, the next work that he did, on the photo-electric effect, he did with the wrong presuppositions, namely, that light is a wave phenomenon which has no particulate component in it at all, no photons. He said in his Nobel prize lecture, 'After ten years of testing and changing and learning . . . this work resulted, *contrary to my expectation* in the first direct experimental proof of Einstein's photon idea.' And that, of course, shows the distinction between those who get hopelessly misled by their presupposition, and those who eventually get led to proper territory.

'Clearly, thematic presuppositions are important for individuals, but do these presuppositions come to dominate a field, I mean, are they shared?'

One must be very careful here because there are great differences between physics and, for example, biology. Within physics there are times when a set of themata is very widely shared, for example the idea that ultimately there will be a unification of the various forces of nature—an astonishing and fruitful idea, and one which a large fraction of the physics community still believes. It isn't that nature told us, in some straight experimental way, 'unification is what you must look for'. You cannot point to meter readings which say, 'here is unity at work.' It is a presupposition that you bring to the field. But it may change in the future.

'Now, there's another phenomenon, in relation to scientists and physics in particular, and that's great schools. Now, it's true that Einstein didn't belong to one, but there have been important groupings of physics at particular times in history which, as it were, come and go. Is there an explanation for that?'

Yes, there is much discussion in the sociology of science on the question of schools, on the question why in some periods there are very many good people in a given field, and at other periods it's rather fallow. One thinks here of Niels Bohr, Heisenberg, Pauli and all the others, all collaborating, all talking to one another at the same time on the frontier problems, and egging one another on to greater heights. Whereas at other times it is very different. I think such concentrations are partly a matter of accident. But partly it is a matter of proper entrepreneurship; that is, schools do not come about by simply waiting for them to aggregate. There usually is behind them a person, or one or two or three persons, who make a decision that this field is worth bringing to the world in an organized way, for example that we need a team. This is how Fermi worked; he put together a (for that time) large group of young people whom he educated and who began a 'school'. This is how Niels Bohr did it, by founding an institute in Copenhagen to which he brought people from all over the world. Rutherford tried the same thing in England. Otto Stern did the same thing in Germany. So these are often decisions made by insightful scientists who take the trouble to find new talent and recruit it to a cause.

'I suppose Fermi's achievement in Italy must stand out amongst that group?'

Well, Fermi, of course, is almost an exemplar of how to produce a school. He started from very unpromising beginnings; that is to say, Italy in the 1920s was extraordinary in some fields, for example, mathematics, but it was bereft of great talent in physics. And so he went on a recruiting spree, and would take a young person about whom he had heard many good things, a 17-year-old, such as Edwardo Amaldi, for example, and say, 'let's go for a bicycle trip to the Dolomites.' During that time he quizzed this chap enough to know that he was a very promising scientist, and added him to his growing group. And the same was true with others. Emilio Segré was also about seventeen or eighteen. Fermi found a piece of rope and wriggled it and said, 'can you tell me why this resonates vigorously in certain spots, and not in others?' From his answers, Fermi discovered that this young man also might be teachable. He did this with an aggregation of people not much younger than himself when he was in his twenties. But it became a real team, very different from the style of Rutherford, who looked for the best person from England, Russia or America who might apply for space in his laboratory. Rutherford had them work separately on different projects that interested him, not together on one big project. But that too, of course, had its own validity.

'But those schools don't last forever, do they?'

They cannot. The whole point about the physical sciences is that a good question is one that you can solve in five to ten years, and it's very different from philosophy, where the good question is one that you can think about forever. But after five or ten years, the frontier minds are likely to find their best place in a new direction. This idea of moving from field to field is different from what one might discover from textbooks. Good minds, and citations to their work, tend to migrate from one area to another as things become much more exciting there. After all, the chief reason for doing science today is to do better science tomorrow.

CARLO RUBBIA
*was born in 1934 and was Director General of CERN
in Geneva, where he is now a Senior Physicist.*

CHAPTER 21

Asking nature

❧

Carlo Rubbia
Particle physicist

MOST people find particle physics very hard. They can find no way to think about the world of the infinitely small, where everyday notions of solidity, visibility, time and space simply do not exist. This is one of the reasons I wanted to talk to the Italian Nobel prize winner, Carlo Rubbia—often called the godfather of particle physics—to try and get a sense of how he thinks.

He was until recently the Director General of CERN, the European laboratory for particle physics in Geneva, and it was here that we interviewed him, shortly before his term of office came to an end. He has spent most of his professional life in these laboratories and it is largely due to him that CERN is now the particle physics capital of the world.

The centre straddles the Swiss–French border. Much of it exists several hundred feet below ground, where a unique array of tunnels and hardware has been constructed to enable beams of matter to be smashed together at such colossal speeds that the collisions give a momentary glimpse of some of the elementary constituents of the universe.

Several thousand scientists of all nationalities do experiments here. It was at CERN back in 1983 that Rubbia and his team made their spectacular discovery of the existence of the W and Z particles which are believed to carry one of the fundamental forces of nature, the so-called weak nuclear force.

Making such a large and disparate group of people work successfully together seems almost as daunting as diving down into the very heart of matter. So, how did Rubbia do it? And what drove him into the strange world of sub-atomic physics in the first place?

———————

I have a family with strong traditions. In particular, my father was an engineer—he had a great respect for applied sciences—and he wanted to see his first son, namely me, become an engineer. So, when I came of the age to go to university, the whole family moved over to Milan so I could go to the best university in Italy—which just happened to be also the university of my father. You see what I mean . . . what kind of thing we are dealing with! But I had a great discomfort with engineering and practical productivity. I really wanted to be driven by curiosity rather than by productivity. So there was a big conflict between my father and me which was handled in a very, very dignified and very proper way—namely, that my father said, 'Well, if you want to study something crazy like physics you can do that, but then you have to support your own studies by yourself', and I said, 'Fine, OK, great, let's do it'.

Now there was in Pisa a sort of college supported fully by the government—in fact it was introduced by Napoleon two centuries before—and it was the only school in which you could have full board, and lots of nice things like that, all paid for by the government, regardless of whether you were poor or rich. This was important because my family was relatively wealthy so I could not enrol for welfare on this. So I tried for a place there—there were ten slots open to study physics—but, believe it or not, I passed eleventh, so I found myself out of this and I had to go back to Milan to learn about slide rules and how to make drawings for engineering and so forth, with great discomfort and great discontent on my side.

Then halfway through the academic year I got a letter because some unknown fellow in that bunch of ten had decided he wanted to quit, so as I was, so to speak, on the waiting list, I got a seat on board. So I was able to move from Milan to Pisa with great pleasure halfway through the academic year just because this mysterious fellow—whom I never met by the way—decided he didn't like science. And that's how I became a physicist.

'There are two puzzles. First of all, how is it that you, one of the world's great physicists, did so badly in that exam?'

Maybe I'm not one of the great physicists.

'Come on.'

The fact was that I wasn't prepared. I didn't do the homework. It was shortly after the war and I had a large multitude of interests. Sometimes I was interested in philosophy . . . some other times I was interested in literature . . . etc. And to pass an exam you've got to do some work—and I was just too superficial, I guess. It wasn't that I wasn't understood, or not fully appreciated—I really flunked the examination, that's a fact.

'Now, you are held in a certain awe in the world of physics, whether you like it or not. There's no question about it. So having, as it were, just scraped into college, when did you realize that you had rather, as it were, special talents?'

I don't believe I have special talents. I have persistence—not professionally but intellectually. I think I picked a number of things I wanted to do and never gave up. After the first failure, second failure, third failure, I kept trying and eventually we got it through. I think that people in science should realize that science is not something that can be done overnight. It's something which is done after a lot of persistent work. And I think the best quality of a true scientist is not sheer intelligence or Albert Einstein's approach, it is just persistence and having no fear of doing work which turns out to be useless.

'You were always keen on physics, were you? It was always physics that attracted you?'

It was the *method* which attracted me—the experimental method, which was born with physics, and is now universal in science. It's asking a question of nature, and listening for the answer from nature. Obviously there are a variety of subjects—some of them are called physics, some of them are called chemistry, some of them are called biology—but there is this common denominator, the way in which you're going about asking the question and detecting the answer. And in my view it's this kind of *method* that attracts me. In a way, I think 'natural philosophy' is the term which really embodies every one of those things. So, really, there is only one type of science and the various fields are somehow different chapters of the same book.

'You've described science as a great adventure, and you've even said that the adventure of doing science is more important than the discovery—that discoveries have really been, perhaps, overemphasized. What did you mean by that?'

Let me put it in this way. I do believe that a scientist is a freelance personality. We're driven by an impulse which is one of curiosity, which is one of the basic instincts that a man has. So we are essentially driven not by . . . how can I say . . . not by the success, but by a sort of passion, namely the desire of understanding better, to possess, if you like, a bigger part of the truth. I do believe that science, for me, is very close to art. Many people think that a scientific discovery is a very rational process—mathematics and equations and so on make it look extremely formal and extremely strict—but in my view, scientific discovery is an irrational act. It's an intuition which turns out to be reality at the end of it—and I see no difference between a scientist developing a marvellous discovery and an artist making a painting, or music or something. Both would be in a certain sense an expression of his or her own self. Our scientific product in general, at least at my level, it's always identified through the person who has created it.

And there's another thing I want to say in this respect, too. The act of discovery, the act of being confronted with a new phenomenon, is a very passionate and very exciting moment in everyone's life. It's the reward for many, many years of effort and, also, of failures. And it's very badly transmitted through the books. A young person who reads a science book is confronted with a number of facts, $x = ma$ . . . $ma = mc^2$ . . . You never see in the scientific books what lies behind the discovery—the struggle and the passion of the person, who made that discovery. I mean, to be a scientist is not a job nine to five. When you do science you have to do science 24 hours a day. When you are at home you should be thinking about science; when you are going to bed, you should be dreaming about science. It's full immersion, you see. Some people ask me, how come you don't have hobbies? The answer is very simple: I don't need hobbies. I mean, why should I run after a ball on a field after I have kicked papers around from nine to five? I am very fortunate that, for me, the interface between a job and pleasure is essentially non-existent. I am enjoying every moment of my work, and my work is my *raison d'être*. I don't see any necessity for compensations of any kind.

'To the outsider, particle physics really does seem very difficult. It would be very helpful to know how you actually *think* in particle physics. What sort of images do you use . . . or is it all mathematical? I have no sense at all of how you do it.'

Well, for me particle physics is a journey . . . it's a journey towards the infinitely small. And this is exactly the opposite of another terrifically exciting journey which is going to the infinitely large, projecting yourself into the universe, going back in time . . . covering big distances. The infinitely small and the infinitely big are equally rich and rewarding adventures. When you dive inside matter it's as exciting as making an infinitely long interplanetary journey. You can see things happening not on a larger scale, but on smaller and smaller scales. You get more and more details, and new pictures come in your mind one after the other.

'But are these pictures in your mind mathematical pictures? When you think, what do you see?'

Well, this is the interesting question. In fact I'd like to elaborate on this. It's just like when you go through a certain process, or go on a journey, you have to accustom yourself to changes. Your mind, your way of looking at things, has to change once we go into the infinitely small, and the argument goes as follows. We normally try to explain ideas of a new nature in terms of something we already know. But if you take a proton, for instance, which is part of the atomic nucleus, and ask 'What is a proton like?', you can't say—I don't know—a proton is like a tennis ball, and so forth, because such a procedure no longer works once you go to the infinitely small. It no longer works because everything changes once you go through the atomic level, which is roughly halfway between the microscopic world and the frontier of our knowledge.

   Let me put it this way . . . we are now down to 10^{-16} cm as the smallest objects that we can perceive through our experiments. At 10^{-8} cm we hit the atomic barrier and that, of course, has created what is called the quantum mechanical revolution and induced very major changes in the way in which we see and perceive and identify experiments. Indeed, quantum mechanics is more a *philosophical* revolution in explaining what you can see and what you can't see. Therefore going through the atomic level is like going through a passage, a gateway. Leaving the world of classical mechanics for quantum mechanics, we have to abandon a number of simple ideas which are part of our everyday experience. We can no longer, when we come down to those things, use simple concepts. As I said, we cannot ask the question 'Is the surface of a proton rough or is it smooth?', because it doesn't make any sense. You can't even ask that question. You cannot ask 'Is the proton red, or is the proton blue . . . does it have a structure, or is it bumpy?' There is no *image*, so to speak, of these things that are infinitely small; our image is so primitive that many of our ideas of vision are no longer valid. For instance, at the sub-atomic level you cannot see the position and the velocity of an object at the same time. It's extremely hard to perceive that. Likewise, relativity has to come in. The particles inside atoms are very small, very tiny, so they travel very fast, and then another kind of a frontier comes in—the speed of light. So once you combine quantum mechanics with relativity, you end up in a terribly complicated and unusual world, which is the world of experimental physics today. So you find physicists talking to each other and saying, you know, this is like that, or like the other, but the comparisons are made in terms of concepts which may sound totally absurd to a normal human being but, in fact, are simply the result of the experience we have built up dealing with things of this type. So, it's a genuine trip you're taking inside matter, inside yourself, inside the very objects which surround us.

'But in a curious way one of your great achievements seems to have been an engineering one, because when you came to discover your famous particles, your W and your Z particles, you actually persuaded CERN to change the way they built the machine.'

Yes, you're right, you're right. You might say I threw engineering out of the door and then it comes back in through the window. But this is the different point, you see. In order to do science, a man always confronts himself with the limitations of his capacity to detect things. Until we had the microscope we could not see living microbes. Until we had the telescope we could not see stars. When we go very deep inside matter the instrumentation becomes the fundamental element, and in parallel with the development of ideas, conceptions, of ways of describing matter, there is a whole chain of marvellous instruments that man has to construct in order to pick up the signals. It's like in space programmes, where many things which a few years earlier you couldn't think of, were not even dreamt or talked about, now become possibilities.

'I still don't understand, though. When you thought about how to detect these particles, how did you come to that idea? I just don't have a sense of the thought or the mental images involved.'

The aim was not a new one. The idea has been a persistent idea for a long time. So the question was not *what* to find but *how* to find it. Therefore the problem was essentially one of inventing a sufficiently clever method which could bring such a thing within reach with the equipment available at a given time.

'So how did you find that method of doing it? That's what I don't understand.'

Well, I would say 'In the privacy of the water closet', but I would not say such a thing to the respectable BBC.

'But maybe it is true.'

Yes. The point is, there *is* a moment when you feel that you want to try the thing, whatever it is. It's an irrational and it's an instinctive moment in which something clicks in your mind and you say, 'Why don't we do this, I mean, why not?' And first of all you take it as a joke initially; you can't afford to take it seriously because you have to be prepared to change your mind. If you are dead serious, it's like having a driver who is afraid of driving—he'll never drive very far or very safely.

So, you start by having something in your mind, and you say, 'Hey, maybe this will work.' If it doesn't, well, you forget about it, and say 'Let's look at something else.' Then two days later you get it again and say, 'Hey, it's still OK, maybe we do have a way of getting through this thing, maybe it *will* work.' Then you get some numbers which look horrible, and say, 'Well, it will not work after all.' But *then* you say, 'Don't give up, just think about it, and maybe if you do this instead of that it will work.' So, you drive your way painfully through a number of different ways . . . it is sort of a game. Any fundamental advances in our field are made by looking at it with a smile of a child who plays a game.

'Now, here at CERN you're rather like an emperor of a small city state.'

I don't think so. You see—you'll be surprised—but in science there is absolutely no authority coming from rank, from position. You don't have to be a boss, you have to be a missionary. You will never get the young scientist to do something unless you *convince* him that he should take care of that. There is no authority in science except the one coming from convincing power.

'But you have a reputation for getting the very best out of people. Is it because they're frightened of you?'

Well, I hope I have a reputation of getting the very best out of *me*, and then I suppose that the rest comes with it. I am honestly speaking, I think that I never try to impose on people more than what I impose upon myself. But my image is the one of a brilliant Italian personality—and I am a man from the South and I am very proud to be a man of the South—so perhaps I am more outspoken. And I do have a threshold for reactions which is probably lower than some other more conscience-driven people. But in a certain sense it comes with my personality. There could be a strong exchange of views, but the day after, it's gone. It's part of an evaporation mechanism. It's not a part of a specific plan.

'But it is amazing that CERN really works. Here you've got these many different nationalities, tremendous co-operation—isn't it a model for how we should all live?'

It sure is. In fact it's been the model for Europe far before Europe was introduced. I mean, CERN was invented thirty-five years ago at a time when Europe was Adenauer, it was De Galle and a few others. CERN is very far from today's European struggle and realities, etc. We are still a product of the romantic phase of Europe and we want to stay that way. In fact, at this present moment, CERN is no longer a European laboratory, it's a world laboratory. If you go around you'll find in our laboratory here that one scientist out of three is non-European.

Americans, Japanese, Russians are coming in by the hundreds to work in our lab, and also Third World countries are now coming in.

'But how do you keep the peace, as it were, between the prima donnas, the different nations, the different demands?'

Let me point out to you that the average age of the scientists working at CERN is about thirty. It's a revolving community of people who are coming in, staying here for five, six years, and then going back and doing something else. So, dealing with such a community is much easier than dealing with the really grown-up and, how can I say, more sophisticated and more mean-looking fellows. In a way, the strength of science comes from the fact that at the moment the highest productivity of scientists is in their early ages, and when people are young they are more tolerant; they are more open to changes. In fact, to be internationalized in science is trivial. The language of science is, by definition, international. There is no British science or Italian science or German science or American science. Everybody contributes. It's a little bit like in sports. The competition among scientists is like competition in sports, and the same spirit you have in the teams, you know, in sports, is also a part of science. But it is also true that even if we are in a totally internationalized, antiseptic sort of atmosphere, individual qualities of individual cultures do come through. Indeed, let me tell you, that here at CERN, and generally in science, in this friendly competition between different people and different countries, the good qualities add up and bad qualities, so to speak, fade away. So I think by mixing people in a wide international environment we produce a better breed of interactions than you would do if you kept all the Italians at one end, all the French at the other, and all the British somewhere else.

'Some people see competition in science as something that's bad.'

Competition in science is a necessity, it's an absolute must. Without competition there would be no science, in the same way that without competition there would be no sport. People who think that they could build a world lab, in which one director would decide on a planetary scale whether certain things should be examined or not, and claim that it would represent a money-saving operation—they're out of their mind! Competition is the fuel of science, of the pride of a scientist; it's the *raison d'être* for many of the scientific activities, and it's a self-regulatory mechanism in which good science gets pushed forward and the bad science gets thrown out. You see, the other point about science that is important is that science cannot be made in isolation. It's impossible for scientists to develop alone. On a desert island there would be no science. And Robinson Crusoe would not get science out of his man Friday, because he has to have a partner, an equal, not a subject. So even on the most marvellous island he would not be a good scientist. Science is done out of debate, out of confrontation; it therefore has to have in it the competition, otherwise it will not fly, it will not be science.

'Well, you're going to step down in about a year's time.'

Yes.

'Will you miss it?'

No, it's my conscious decision. I feel that to be a director of a laboratory should not be, by definition, a permanent mission. People should have the courage to step down and go back to science. I believe you will never have a good director of a scientific laboratory unless that director knows he is prepared to become a scientist again. And not only knows it but means it. And in this respect I feel I should say that I gave my contribution; I spent five years of my life to work hard for other people's interest. It was not easy, I must say that every one of those years counts twice, as far as, you know, the work and the consumption and the effort it has taken, and I do believe that now, for me, it's time to go back to science again. I have some wonderful ideas, I feel I'm re-born, and I hope that this will come very soon.

DAVID PILBEAM
was born in 1940 and is Henry Ford II Professor of the
Social Sciences at Harvard University.

CHAPTER 22
A very tidy desk

❧⸻❧

David Pilbeam
Physical anthropologist

THIS is the age of big science; high-tech science. Of particle physicists using vast machines to mimic the very act of Creation; of molecular biologists unravelling the secrets of DNA, from the mutations which cause inherited disease to the relentless clock of random genetic change which has been ticking away, independent of environmental pressures, since life began. Where, I wanted to ask David Pilbeam, did the study of human evolution fit into a picture like this?

The popular image of palaeoanthropology is still one of charismatic, driven figures, combing the Earth's wild places for fossils, the few mute fragments of teeth and bone which are all that remain of our remote past; of tentative and even rather romantic accounts of humanity struggling painfully to be born. Where is the precision, the wealth of data, even the ability to re-run the experiment that modern science takes for granted? How do biologists go about the seemingly impossible task of reconstructing human origins?

────────

First of all, it is not a science, 'it' being palaeoanthropology, the study of evolution. It's not a science in itself, it's a question: what happened in human evolution and why? And in order to answer the question one has to use many different kinds of information. In fact I don't think it's true to say that palaeoanthropology is just going out and getting fossils, or mostly going out and getting fossils, because right from the start of picking up a fossil you can't make any sense at all out of it unless you have an understanding of anatomy, particularly comparative anatomy, unless you also have some understanding of functional anatomy, behavioural ecology, the ecological context, and so on. These are not facts that speak for themselves, they're facts that you have to put a lot into.

'But your raw material is the fossil?'

My raw material is mostly fossils but, of course, it was perfectly possible for Darwin to speculate about human evolution quite constructively without a fossil record, and so it's possible to—in fact, lots of people do—talk about human evolution paying very little attention to the fossil record. And it's my own view that a very great deal of what palaeontologists do when they talk about the really good stuff, you know, the blood and sex and thunder and aggression, has got very little to do with the fossil record, and has lots to do with all kinds of other records.

'But you said it's not science.'

It's not a science in the sense that physics is a science. Or that chemistry is a science—if you're a chemist, that is. If you're a physicist, you probably wouldn't say that chemistry is a science! But human evolution is not a science in the sense that physics is—in having its very well delimited notion of theory, and fundamental agreement about what the important issues and important questions are. Palaeoanthropology has no theory of its own: it's part of evolutionary biology. It's a small corner of evolutionary biology where we're working on a not very diverse group of organisms which would be, if they weren't studied by us, really rather dull. I mean, if we were pigs working on evolutionary biology we would find suid evolution much more interesting than human evolution.

'Well, that is one of the points that puzzles me about it, because it seems to me that no one would really choose, if they wanted to understand basic scientific questions, to study man, or his evolution, because the material is so poor. For example, I'm an embryologist, and I wouldn't dream of trying to work on human embryos in order to understand fundamental mechanisms. It's ethically not possible, but it's also such difficult material to handle. Yet you have chosen precisely this "unsatisfactory", if you like, human subject matter.'

That's why I say that it's the *question* that's the issue—what happened in human evolution and why. And the way you go about answering it is to use the basic principles of biology, in particular, evolutionary biology, that can only have been worked out by studying other organisms. Now, J. Z. Young wrote in one of his books on human biology that human biology is as good a topic for the study of certain kinds of biology as any other, and in some ways that's true, but in the main I would agree with you. There are very, very few examples that I can think of where interesting, broad, general principles come out of the study of humans or, indeed, out of the study of any single group. That's not how you come up with broad principles. The whole point of science is to study lots of cases, try to abstract and try to understand what the fundamental mechanisms are. You can't get at that by studying what are essentially unique circumstances.

'So, why are you studying human evolution?'

Why does one, or why do I personally? I, personally, because I would like to know what happened and why. I think that's a very interesting question and one happens to be able to make a reasonable living trying to answer it, and I think that's actually why most of us who are palaeoanthropologists of one kind or another are studying human evolution. We're, in a way, sorts of historians. Why do historians study history? Because they really want to understand how the present came to be. They would probably say that they aren't studying that, but I suspect that's what motivates most of them. They're interested in the human condition. There's a kind of cultural imperative. Our culture is very much oriented towards that. You strike very sympathetic chords in people all the time when you talk about human evolution, human origins, the present, the future. My colleague Matt Cartmill says that it's because we have become increasingly aware of how dangerous we are to ourselves and to others, and when that sort of thing happens, one way you deal with it is through myth, through the use of mythical stories. Now, of course, myths don't have to be fictional, or don't have to be entirely fictional, and one of the things palaeoanthropologists have done is to serve as myth-makers for the secular world.

'So, there's an emotional commitment?'

Oh I think very much so, and one of the interesting things that's happened in the field over the last ten years is the recognition that one has to be really quite reflective to do palaeoanthropology well. A good deal of what creeps in, in the way of not just explanation, but also in the description of what happened, reflects not so much what is written there in the record, whether it's the fossil record or the genetic record, but also what's imposed upon the record by us as humans, what we bring to the study of the past. Very often the past has been studied—and this applies equally to history—the past has been studied to legitimize some version of the present; whether it's to legitimize the notion that humans are inherently aggressive, and therefore we discover evidence for the aggressiveness of our ancestors; or that humans are inherently nice, co-operative, cuddly creatures if you just give them half a chance, and therefore we can find evidence for that in the record. And you know many other examples that I could think of.

'Well, as I understand it, Misia Landau has actually suggested that what your subject does is to create just those sort of folk tales.'

She's suggested that, certainly well into the 1950s, and maybe the 1960s, major accounts of human evolution are basically hero stories. They describe how initially this kind of barely formed creature begins to venture out and is starting to become human, and is faced with various kinds of challenges which sometimes are met immediately, or sometimes are only met after several attempts, and finally emerges triumphant. I think this kind of view of human evolution is actually quite commonly held, not just professionally, but by all kinds of people. Many of us, I'm sure, have seen the film *Planet of the Apes*, where something happens to humans in the past and chimpanzees and gorillas and orangs evolve into various kinds of human beings. Implicit in the original book and in the film is the notion that somehow everything is striving to become human, that being human is inevitable. It's analogous to the Whig view of history, the progressive notion of the human condition, that everything is very good in the present, and that everything in the past has been striving to become the present. If you come to the human story with that implicit agenda, then you can almost certainly read that into it.

'You're making it sound more and more like history, almost like an arts subject, and the thought that strikes me is that perhaps that's an attractive feature because it offers the scientist, or the palaeontologist, a sort of creative role, a much more creative role than perhaps in other sciences where he or she is much more constrained.'

Well, first of all let me say that I think that in the last twenty years the situation has changed somewhat. Many more of us have become aware of what goes on and are being far more cautious. We are now producing much more boring stories about human evolution; stories that much less resemble hero stories; that have much less mythic or symbolic content to them.

Some of us get a lot of mileage, or have got a lot of mileage, out of selling the idea that only by really understanding the human past can we understand our present condition and say something about the future. And what is interesting to me is not so much whether that's true or not—I suspect that down to really quite deep levels it's not true—but why that approach should be so enthusiastically welcomed by so many people.

'Do you really believe that understanding your subject would be helpful in that way?'

No. I think that some of my colleagues probably do, but I don't, no. I think at one stage that I did, but I no longer do. Except in one way. I've found that trying to understand and to tease apart in my own mind the interplay between what I would call scientific and ideological factors has been very good for me in clarifying my own thinking, and also in clarifying a great deal of what other people have written about human evolution and the human condition. And I see very frequently, now, that it's really like looking in a mirror and seeing a familiar face. Much of what is written about the past actually reveals what is happening in the present.

'Does this mean that it's difficult to falsify things in your subject, or verify them?'

What most of us would call the really interesting hypotheses—if that's not dignifying them too much—are very difficult to falsify. The things that are relatively easy to falsify are probably relatively trivial. What I think the important thing to do in the field is constantly to be pushing to try to expand those testable questions beyond the trivial into the interesting. I'll give you an example of what I think is a potentially very interesting line of work. It matters whether our ancestors of 2, 3 and 4 million years ago—the australopithecines, these small-brained bipedal creatures—whether they matured slowly like us, or more rapidly like apes, because it has an impact on their potential for learning and cultural behaviour and all that kind of thing. Until relatively recently there's been really no way of getting a handle on maturation rates in fossils, but very recently some relatively smart people have begun to look at things like the histology—the fine-scale anatomy—of teeth in much closer detail, and have been able to figure out what seem to be fairly standard patterns of enamel growth, that seem to hold up across many different species of mammals. Now, there seems to be a fairly good relationship between certain patterns and rates of enamel deposition, so you can use those patterns to estimate how long it took teeth to form in australopithecines and how old they were when they died. And it turns out that they didn't grow up slowly like humans, as many people had thought, but they grew up much more like apes. Now from the point of view of a behavioural biologist, who also watches chimps or baboons, that's actually quite an interesting piece of information. So the trick is to figure out how you can extract more information that is biologically interesting and relevant from the material that we have.

'What's your particular skill?'

I think that my strength is being able to make the kind of synthesis that occasionally opens up new questions and new ways of thinking about things. Putting material, diverse material, together in ways that haven't been done before. What I most enjoy about the field is its eclectic nature. The fact that I can be a little bit of a molecular biologist, in the sense of being interested in genetic evolution, and a little bit of a behavioural biologist, and a little bit of a functional anatomist, and a bit of a historian.

'Have you ever been wrong?'

Many times, yes, many times.

'Many times?'

Yes, yes, yes. I think all the time one is trying to assault ones ideas, and that's what I try to do all the time.

'Did you mind being wrong?'

As far as my colleagues are concerned, as far as the outside world is concerned, I've been monumentally wrong once.

'Monumentally wrong?'

Yes, I mean substantially wrong, once. And no, I didn't mind.

'Why were you wrong—in retrospect? Hindsight's wonderful.'

Several reasons. What I was wrong about—if I can take it a little bit at a time. I was wrong about the nature of the very earliest hominids, about how old they were, first of all. I believed, along with many other people, but that does not absolve me, I believed that hominids evolved quite early on, 15, 20 million years ago. We would now think that's wrong by at least a factor of two, that it's only 5 to 7 million years ago. And I also believed that a particular fossil hominoid called *Ramapithecus* was an early hominid. Now, the reason that I was wrong is, first of all, that we are now able to show from studying patterns of genetic difference among living species that it's extremely unlikely that humans and chimps and gorillas have a common ancestor that's any older than about 6 or 7 million years ago. We can do this because we now know genetic change ticks away like a clock. We can tell how long it took to produce the genetic difference between us and our common ancestor. Also, we now know from a much better fossil record than we had at the time, that *Ramapithecus* is actually not a different species at all. Its the female of something called *Sivapithecus*, and that *Sivapithecus* is probably ancestral to the living Asian ape, the orang-utan.

'So, you were wrong, not because you made an error, but from the available data you made a judgement?'

I made a judgement, yes, but first of all I had made the decision to exclude the genetic information. One of the reasons I did that was because I, along with almost everybody else in my generation, grew up with natural selection as the main organizing paradigm for producing diversity in the evolutionary record, that natural selection played the major role in shaping the way a species evolved. Therefore we didn't believe that genetic patterns could work as a kind of clock. It couldn't be true! And when people came along and said, 'Well, if you look at the genetic patterns in the right way, you can infer times of divergence of past species just from looking at patterns of difference in the living species', I and many others said, 'Nonsense'. It wasn't until 1967 that Kimura came along with his neutral theory of molecular evolution and began to point out that a great deal of evolution at the level of nucleotides and proteins is apparently not affected by natural selection at all, or is affected only minimally.

So my first mistake was to reject the molecular clock because it made no sense, didn't fit into my way of thinking. What I should have done is to look at the data that had been produced, because when you actually look at the data that were produced, even back in the 1960s, they're not as tidy as one would like, but certainly provide very strong support for the notion that certain parts of the genome are evolving essentially at constant rates.

'How did you come into human evolution?'

I went up to Cambridge as a medical student, and didn't enjoy preclinical medicine very much, and was probably correct in feeling that I probably wouldn't make a very good doctor, and so I decided not to be a doctor. I finished the preclinical phase of my education and then I read what was called 'physical anthropology'. I had no plans to do it professionally. I really backed into academia. The man I worked with at Cambridge had spent some time in the States before the war, and he said one day, 'Why don't you go to the States, there's a young fellow named Simons [q.v.] working at Yale on a topic you're interested in, why don't you go and spend a year there?' So I said, 'Yes'.

'Did Elwyn Simons have an important influence on you?'
Yes he did.

'In what way?'
Well, he's a man who's totally committed to and obsessed by fossils. He loves

them, he loves to collect them. He's a man of monomaniacal enthusiasm, and it was wonderful being swept up in that kind of enthusiasm. I felt like a kind of *aide-de-camp* to a great general, and he engaged me immediately in his own research right from the start.

'Which was human evolution?'

Yes, he was very much interested in the very early phases of human evolution, and we began to write papers together and that sort of thing. And I loved it. It was a wonderful experience.

'Clearly my image of your subject is rather outdated, and it has changed enormously in recent years. Is that true?'

Yes I think so. I think that the sea change has taken place within the span of my career, but particularly in the last fifteen years. You could caricature it roughly as follows. Thirty years or so ago there weren't that many fossils. There were not that many practitioners of the subject either, and the proportion of big men to the rest was really quite high—big men being people who went out and found the material, made a big fuss about it, had steel egos and all the rest. It was also the case that notions about the course of human evolution were really quite simple-minded—that there weren't many side branches, that it was a pretty straightforward and progressive process. I think it clearly is the case that accounts of human evolution did resemble hero stories. Things are so different now. We have many, many, many more fossils. The field itself is much bigger. There are lots and lots of younger people in it and the overall level of technical competence and training is really very high. But field work is still terribly important. The raw material that one works on are the fossils. I think it's still the case that doing field work is, in a sense, establishing your credentials. I enjoyed much of my early field work, but I don't want to go off two and three months at a time any more.

'Is it expected of you?'

I don't think so, no, I don't think it is any more. There are many more things to do. The subject has emerged as a much more eclectic, puzzle-like process. I said early on that it's not a science in that it doesn't have the unity of physics, but it's a challenge. You're trying to obey the rules of the scientific game in answering this question, and trying to be as honest as possible. You're pulling in data from all different kinds of directions, but the target's always moving, in a sense. A friend of mine who's a palaeoanthropologist, and also a medical doctor, said that he enjoyed it when at least once a week he could do some simple surgery—because at the end of the day he could go home and have a martini and feel that he'd finished something, it was done. If you do palaeoanthropology it's almost never done, or if you've figured out something, then that bit is no longer interesting, and you go on to the next thing. It's a kind of jigsaw puzzle with 1000 pieces and someone's taken 500 of them away.

'So, in a way you've got to be highly ingenious and far-ranging?'

Very much so, and you also have to be able, as Richard Feynman, the late, great physicist said, you have to be able to live with doubt and uncertainty.

'Do you mind that?'

No, I think that's why I do what I do. I enjoy it very much. But it's also one of the reasons why in the rest of my life I'm one of the most compulsive human beings that anybody has ever met.

'Compulsive?'

Keep a very, very tidy desk.

'That's where you create your order because you can't really do it in your subject?'

No. It's about order which I strive constantly to create in my subject, and it's a subject where you can get some order, but it might not be as much as you would ideally like. In fact, it rarely is as much as you would ideally like.

'And, given what one knows about fossils, it's always going to remain incomplete?'

Yes. It will remain incomplete, and by 'it' I mean the picture that you would like to paint of a species if it were alive today. If you were sitting out in the field studying baboons, or chimpanzees, or hyenas, you have a kind of checklist that you'd run down in order to fill out a picture so that you could write your book if you're Jane Goodall or someone like that. That's exactly the same checklist that I think we should bring to the past, because they're asking the biological questions, and those are the interesting ones. Some of them are probably going always to be unanswerable but, as I said earlier, the trick is that the smart people are those who can figure out ways of answering questions that we previously thought were unanswerable. We previously thought that you couldn't figure out what maturation rates were.

'But in a way you're in a subject which will always remain incomplete?'

Yes, yes. It's very like impressionist paintings. You can't afford to get too close to them because they just dissolve into meaningless blobs. You have to stand back, and then you'll get an impression. And I think the good evolutionary biologist, the good palaeoanthropologist, is someone who's a good impressionist; who's going to be able to paint, sketch, a quick picture that's got the essence, that captures the critical features.

ANN MCLAREN
*was born in 1927 and was Director of the MRC
Mammalian Development Unit. She is Principal
Research Associate at Wellcome/CRC Institute,
Cambridge.*

CHAPTER 23

Of mice and mothers

⟅⟩⟅⟩

Anne McLaren

Mammalian developmental biologist

THERE are remarkably few outstanding women scientists. Anne McLaren is one of a small group which forms the exception. She went up to Oxford just after the war to study zoology, and became interested in the complex interface between genetics and reproduction. In the early fifties she and her then colleague, Donald Michie, were among the first to use the technique of embryo transfer in animal experiments, and were associated with the early research leading to *in vitro* fertilization. More recently she has been working on sex determination in mice, and in a series of very elegant experiments, carried out in collaboration with Elizabeth Simpson, showed that the favoured view that maleness resided on a particular part of the Y chromosome was wrong. Her work for the Agricultural Research Council, and as Director of the Mammalian Development Unit of the Medical Research Council has played a major role in making Britain pre-eminent in the area of mammalian developmental biology.

As she is one of a select group of women in the upper echelons of British science, I wanted to know what her experience had been. Did the male majority block women's progress? Were women put off continuing in science by the conflicting demands of an ambitious research career, and a family? Why, in her view, were there so few women like herself?

———

I think there are a number of reasons. One is that it's a cohort phenomenon in the sense that when one talks about 'successful scientists' today, one is thinking of a certain age group. And if you go back the several decades to when they started out, it was at a time when it was much less common for women to get in to science. Another is that there are social difficulties for women in science if they want to get married and have a family. Child-care arrangements are difficult. All that side of things discriminates against women. I think there are some fields of science where, certainly in the past in this country, there has been actual discrimination against women in jobs. I've been lucky for the whole of my working life. In biology, there hasn't been any discrimination against women that I've ever come across. It's been different in America, but not in this country.

'But what about women today. Do you think that things have changed? Is it still the women not going in to science, or that they go in, but don't survive as well?'

In my field they go in, and they do survive rather well. In my field a lot of the leading scientists now are women, and my guess is that in twenty years' time one will see quite a change there.

'But if we actually look at University College where you and I work, there are very few women professors in general.'

What about lecturers, though. Because when we say that things are changing *now*, one ought to look at the lectureship level, shouldn't one, rather than the professors who are the earlier cohort?

'I suppose so.'

And the other thing is that I don't think that women are as keen to be professors as men are. I think they're very interested in research, and they're interested in teaching, but I have known men who were really very keen to be professors. I remember once congratulating somebody on having got a Chair, and he beamed, and said, 'yes it's been my ambition to become a professor ever since I was a little boy', which I thought was nice.

'You think no woman could say that?'

No, I'm sure some could. I just haven't met any.

'How then did you, belonging to that small, early cohort, choose to go into science?'

Rather by accident really. I'd decided that I wanted to go to a university, which was a new thing for women in my family, and it was suggested that I should read English, because that was a suitable thing for women to read. Also, I was good at writing essays, and so in the little schooling I'd had, English always came out rather well. This was during the war when the whole schooling business was very difficult. I didn't go to school, in fact. But I was going to go to some ladylike 'crammer' in the country to study for Oxford entrance. And before I went I actually looked up some of the entrance papers, and discovered that to do the entrance exam in English Literature you had to have read an awful lot of books by people like Dickens, and poems by people like Milton. I never had, and I realized that it was not really 'on' in the eight months I had available to study for the exam. So I looked through the other entrance papers, and found that zoology seemed to be something where one could probably answer the questions with minimum swotting up. So that's what I settled for, but I don't regret it for a moment.

'It's still not clear to me why you actually chose to go to university?'

Well I don't know really, but I guess during the war people were thinking in new ways, and deciding what they wanted to do. The alternative, in my family, would have been something that in those days was called 'coming out' which meant going to a lot of dances and parties, and meeting people, and that didn't seem to me a very attractive option. And anyway, I'd always been interested in science. When I was a child I liked trying to make things. I never really had the proper training or equipment. I don't know whether you've ever tried carving cog wheels out of cork, for instance, but it's not very satisfactory. But at least I tried.

'So you never really set out to be a scientist?'

No, no.

'What was Oxford like when you went up there, that was in 1945?'

Enormous fun because it was when everybody was getting demobbed and coming back from the war. So most people there were much more mature, and had much more experience of life than one would get in a normal first year. And also it was a revelation to me that people thought and talked about so many interesting things, and did so many interesting things. It was wonderful.

'When Dorothy Hodgkin was at Oxford she was actually excluded from one of the chemical societies, the Alembic Club, simply because she was a woman. Now you were there much later, did you find any such discrimination?'

No, none at all. Of course, I didn't do chemistry. When I was doing three subjects, zoology, physics and maths, it was quite interesting because in zoology there were almost as many women in the class as men. In maths, I would say, perhaps between 10 and 20 per cent of the class were women. But in physics, I was the only woman, and it may be that if I had gone on with physics later I would have found myself excluded from certain clubs, I don't know, but in biology certainly not.

'Was zoology quite old-fashioned still?'

I learnt about animals, if that's what you mean; but I think that's quite a good thing in zoology. Then in addition to learning about the animal kingdom there were excellent series of lectures on genetics and evolution. E. B. Ford was there and taught genetics, and other wider aspects. Of course, there wasn't molecular biology, but there was a lot of interest.

'When did you decide that you actually wanted to do research?'

I suppose while I was an undergraduate. I got more and more interested in zoology. There seemed more and more interesting things to investigate, and I wanted to investigate them.

'What subject did you choose for your Ph.D.?'

I started off on what I thought was a very interesting semi-genetical subject. There had been a report in the literature that if you immunize pregnant rabbits against lens protein, the young get eye defects. It was only a short paper, but I thought it would be worth repeating, and seeing if one could understand the mechanism. The paper also claimed that the eye defects were inherited. Now, unfortunately, none of the rabbits in the Oxford animal house got pregnant in the whole of the first year that I was supposed to be doing my Ph.D. so that didn't seem a very promising start. But I learnt a lot of immunology techniques, and actually somebody later did show that the eye defects were genuine. But I switched to working on neurotropic viruses—it was the days when polio was still a big problem, and there were model mouse viruses—because viruses reproduce quickly, and were an easy subject to get a D.Phil. on in two years.

'But then you changed.'

Then I went back to what I was interested in before, genetics, maternal effects. I was always interested in the perhaps less conventional sorts of genetics, and the next problem I looked at was a maternal effect. There were two strains of mice which differed in the number of lumbar vertebrae that they had in their back, you know, between the ribs and the pelvis; and if you took a female from one and crossed it with a male of the other, the young resembled their mother and not their father, and this seemed to me interesting. This was collaborative work that I was doing with Donald Michie at the time, and we decided to try to see whether the maternal effect was being exerted through the cytoplasm of the egg because, of course, the mother contributes much more cytoplasm in her egg than the father does in his sperm, or whether it was a uterine effect, something that happened in the mother's uterus after implantation. And we used the fairly recently introduced technique of embryo transfer, taking embryos out of one female, and putting them into the uterus of a female of the other strain, and, in fact, it turned out that it was a uterine effect and not a cytoplasmic one. The young resembled their foster mothers, their uterine foster mothers, and not their real genetic mothers. And as far as I know, that's the only well-documented case of an actual morphological characteristic which is determined by the type of uterus that the young are gestated in.

'What was the general problem that you were interested in?'

Genetics, development, heredity, reproduction. I think the borderline between genetics and developmental biology. And always in mammals. I think I'm interested in mammals because we're mammals, and if you're interested in developmental biology and genetics in mammals, you also have to be interested in reproductive biology.

'But that's the question I want to ask you. Was there some grand problem?'

No, no grand problem. Opportunistic. Whatever seemed most interesting, and most promising at the time.

'So it was really just hunting around, or keeping your eyes open?'

Not so much hunting around, but selecting from the really large number of interesting problems that are always cropping up, and trying not to get distracted on to a new one before one had got somewhere with the last one.

'That might be thought of as a strange way of doing research. I know it's been enormously successful in your case, but essentially you chose *not* a specific problem but . . .'

. . . a field.

'You chose a field and *then* saw where you could make advances? Would that be a right way of describing it?'

Yes. And I think connected with that is the fact that virtually all my life I've worked on the mouse as a model system. I'm not particularly fascinated by how mice reproduce and develop, but for fifteen years I worked for the Agricultural Research Council, and they presumably saw my mice as little model pigs and sheep, and since I've been working for the Medical Research Council, where, of course, it's been more human and clinical problems that's been at the back of one's mind. But I don't think I've ever had any grand strategy.

'Do you actually like your mice?'

Very much, yes. I love my mice.

'Do you handle them yourself?'

Yes, yes.

'Do you like doing the experiments with your own hands?'

Yes I do.

'Do you think that's important?'

I think different people do science in different ways for different motivations. It's important for me, but I know a lot of people don't like bench work, and they're very good at running a whole team of assistants who do the bench work. I would find it much more difficult to work that way.

'Do you actually spend a lot of time handling the mice and breeding them yourself?'

Yes. And, of course, one notices things about them which one wouldn't notice if somebody else was doing it. So it's useful from that point of view.

'What pleasure do you actually get from doing research?'

I think the bit I enjoy most is analysing data. If one is presented with a pile of raw data, and can turn it into a satisfying story, that's very enjoyable. I think that's what I enjoy most.

'Do you actually enjoy communicating that?'

No, no, no. I communicate because I have to, both orally and in writing, but it's not a side of science that I enjoy.

'But when you've made a discovery do you not want to tell everyone about it?'

No, no.

'Do you mean it's just for you?'

Maybe the people in the lab; if I was excited I'd go and tell them. And I think it's important, very important, that one should communicate one's science as widely as possible, both to other scientists, and to the public. I always reckon that writing papers, writing articles, and talking about one's work is at least as important as the actual lab research side, because there's no point finding something out if you keep it to yourself. It's just that I don't actually enjoy it.

'Do you not want in any way one's peer groups approval then?'

Well I think that's the people in the lab, and yes, other people working in the same field, one discusses things with them; and if one has an interesting result that's exciting, and they tell you their interesting results and, of course, I like that sort of scientific discourse; but not the honour and the glory side, that doesn't appeal to me.

'You don't sound very competitive .'

Well some people would be surprised to hear you say that, but no, I don't think I'm very competitive. It doesn't worry me very much if somebody gets a result which I might have got given another few months' work, providing I've got other things in the pipeline. Of course, if it was the only thing I'd been working on for two or three years I'd be put out, but otherwise it wouldn't worry me.

'Do you think that's a female quality? I think there are a lot of men in science who are very competitive, very keen to publish rapidly, and to get there first.'

I think there are a lot of women in science of whom the same could be said. No, I don't think that is a specifically female or male quality.

'So that's a peculiarity to you?'

Yes.

'In running your life, how have you managed to combine your science with your family, raising children?'

I guess by not being a perfectionist. I'd have been a better scientist if I hadn't had a family, and I would have been a better parent if I hadn't been a full-time scientist, but I've enjoyed both sides of my life very much; am still enjoying them very much.

'So you think it really is a disadvantage then, both as a mother and as a scientist, to bring up a family and do science?'

It's certainly a disadvantage timewise. I'm not sure about other ways. Again I think it depends on the person. My children say that they're very glad I was a working mother, because it made family life more interesting, and I think if I'd spent all my time with my children when they were small I might actually not get on so well with them now. So I think there are pros and cons.

'Do you think there is among women scientists a problem of guilt, of not spending enough time with their children?'

Yes I think there very often is. I'm lucky, I don't suffer from guilt, and so that hasn't worried me, but I've certainly seen it in other people.

'When you say you don't suffer from guilt, is that a general statement?'

Yes.

'How does that come about?'

How do people's characteristics come about? It seems to me a thoroughly unconstructive sort of feeling, and I reckon that I'm very lucky in not being affected by it more than I am.

'Are you very single-minded?'

No, dreadfully *un*-single-minded. I try and do far too many things at the same time. I think probably any woman who has had a job and brought up a family of children, can't afford to be single-minded because one is always thinking of half a dozen things at the same time. I greatly respect and envy my scientific colleagues who are single-minded, but I've never been single-minded.

'Are you obsessional at all?'

Yes, I think I am obsessional. I think I'm prepared to go on banging my head against a brick wall for longer than it would be sensible to do so.

'So you're obsessional as distinct from being a perfectionist?'

Yes, I think they're quite different.

'In what way?'

There's nothing perfectionist in going on banging one's head against a brick wall, it's just obsessional.

'But I thought the mark of a really great scientist, or a distinguished scientist, which you are, is knowing when to stop banging your head against a brick wall.'

Yes, yes, and fortunately I'm not so obsessional that I go on until my head drops off. But perfectionism is actually much more harmful, because if you're a perfectionist in science you never get anything finished and written up and so move on to the next thing, because no experiment is ever the last experiment, and the last word. There are always unfinished bits and pieces, you always feel you could do it better, but it's really very unhelpful to be too perfectionist.

'When you started with mammals, and on your mice, did you anticipate in any way at all the whole question of experiments on human embryos or *in vitro* fertilization?'

Oh very much so, yes. When Donald and I were doing embryo transfers back in the fifties, it was the time when I was having children too, and therefore my friends were all in the reproductive age group. And I remember going round and asking people which would they feel was more their own, a baby that they were carrying, even though the egg had been provided by another woman, or a baby that was born from their own egg which had been gestated by another woman. One didn't in those days use the term surrogate mother, and most women that I talked to said they'd feel it was more their own if they'd given birth to it, whether or not it had been their egg, and I felt the same. No, one was very conscious that this technique would be applied sooner or later for clinical purposes.

'And this was before the work of Edwards and Steptoe?'

Bob Edwards was a contemporary of mine. He started off working on mice and rabbits, and then he moved over into the human side, but we kept very closely in touch, and I like to think that some of what I did fed in to his eventual achievements.

'Since you were working in the field, was it from the beginning the idea that one would try and do *in vitro* fertilization in humans?'

At the beginning one didn't realize that it would have to be *in vitro* fertilization, as well as embryo transfer. I think initially I was thinking more in terms of women, who for one reason or another, couldn't carry a pregnancy to full term, but could produce an egg and have it fertilized *in vivo*. That would be the situation where one would want to do embryo transfer. It really wasn't until Patrick Steptoe came on the scene, and he and Bob Edwards were able to recover eggs from the human ovary, that the *in vitro* fertilization side of it came in.

'Did you anticipate any problems at that early stage? You'd obviously been around asking people, so you perhaps thought there were going to be problems?'

Yes I think so, if one goes back and reads what one wrote. I mean not only what I wrote, but also what Bob Edwards wrote, and other people round about 1970, say. All the problems really are anticipated there, but people didn't take it seriously, and I remember we tried to get the Developmental Biology Society, or other societies, to have sessions where they would discuss possible social and ethical implications, but people, including lawyers and ethicists, tended to brush it off.

'What was your main concern?'

I think my main concern was that people would be so put off by the scary implications of that sort of reproductive technology that they wouldn't appreciate the very real benefits for human welfare and women's fertility. There was a tendency in those days to talk about *in vitro* fertilization as though it would produce Frankensteins. It didn't and, in fact, by the late seventies, by the time Louise Brown, the first 'test-tube' baby, was born, the climate of public opinion had changed. The newspaper treatment of Louise Brown's birth was very, very different, and much more optimistic, than what the newspapers had been saying in the late sixties and early seventies.

'You had a very important influence in setting the limit up to which people could reasonably work on human embryos, at fourteen days after conception. How did you arrive at that?'

By looking at the course of human development. Seeing where the possible landmarks are that one could use as the place to draw the line, and deciding which of them perhaps was the most appropriate. But it was the Warnock Committee as a whole, it wasn't just me.

'Do you think the Warnock Committee, on which you served, was a success in the sense that that's the right way to go about dealing with a problem like this?'

Yes I do, and I think it's been the envy of many other countries. When one travels around the world and discusses these matters, almost everybody is very impressed by the Warnock Committee—the way in which the whole question was handled, and the amount of public debate that there's been in this country since. Because in other countries, in the main, either very little has been done, or legislation has been pushed through rather hurriedly, without much opportunity for public discussion and public education.

'Where do you feel, or how do you think one should handle science in the public domain? I was struck, for example, that when the AIDS scare, or the AIDS epidemic started, you wrote a very sensible letter to *Nature* suggesting using condoms. Why did you do that?'

I'd like to see things changed for the better. I'd like to see society changed for the better. I'd like to see science used to help people, and anything that I can do towards that end I will try to do, though usually rather ineffectively.

'So that does mean that you do care about the use of science?'

Very much, very much.

'Applied research is therefore important to you in a way that, perhaps to many other scientists it isn't?'

Yes. I've never actually done any of what is normally called applied research, but all the research that I have done has been in the area of strategic research where I've had applications in mind, although I haven't been directly working on them. I'd have been quite happy to work on applied research, it's just never come my way.

'A lot of your research is, as you've said, strategic in relation to human welfare. Is that your main motivation for research?'

It's certainly a strong motivation. I don't think I'd say that it was my main motivation, because I think even if I was stuck away somewhere all on my own, without any motivation of that sort, I would still want to try to understand what was going on around me.

AVRION MITCHISON
was born in 1928 and was Professor of Zoology at
University College, London. He is now Director of the
Deutsches Rheuma Forschungs Zentrum in Berlin.

CHAPTER 24

A family affair

❧

Avrion Mitchison
Immunologist

CERTAIN families seem to breed great biologists. Darwin's grandfather, Erasmus Darwin, was, in his own time, equally as famous as his grandson. Thomas Henry Huxley, Darwin's defender, was President of the Royal Society, and so was his grandson, the Nobel laureate, Andrew Huxley. No less distinguished is the line descended from the physiologist, J. S. Haldane. His son, J. B. S. Haldane, was a founder of biochemical and population genetics, and became professor of genetics at University College, London. Haldane's sister is the writer, Naomi Mitchison, and Avrion Mitchison is the youngest of her three sons. He is a leading immunologist and, at the time we spoke to him, was Professor of Zoology at University College, London. Murdoch, his older brother, is Professor of Zoology in Edinburgh, and the oldest brother, Dennis, recently retired as Professor of Bacteriology at the Royal Postgraduate Medical School in London.

What factors make for such a strong family tradition? How great is the influence of what one might call the English country house tradition, that particular combination, in some houses at least, of deep respect for both the intellectual life and the surroundings of the natural world?

Mitchison went up to Oxford to study zoology just as the science of immunology was beginning to develop. Here he was taught by Peter Medawar whose pioneering work on the immune system was to lead to a Nobel prize. Medawar was strikingly good looking, a brilliant lecturer and writer, and his influence on Mitchison was profound. Medawar, too, became Professor of Zoology at University College.

Medawar's good looks and elegance could be intimidating. Mitchison's style is much warmer. His informality of dress and manner sets him apart from many other university professors, and he's known both for his directness and his kindness. But he also has a reputation for intellectual ruthlessness, a quality for which his uncle, J. B. S. Haldane, was noted. How much, I asked him, had being a Haldane determined his choice of career?

———

Genetically I shouldn't think at all. Culturally, certainly very strongly. It never crossed my mind, as far as I can remember, to engage in anything except the practice of science. I was brought up, I suppose mainly by my mother, deeply to respect her brother, J. B. S. Haldane, and indeed, also deeply to respect my grandfather, whom I can only just remember, the physiologist, J. S. Haldane.

'But did Haldane himself have a great deal of influence on you?'

Yes. Like, I suppose, lots of growing lads with a relation who's revered in their part of the family, I desperately hoped that he would take an interest in me, and I badgered him from time to time in the hope of him doing so. I wasn't actively rebuffed, and so the admiration and interest burgeoned.

'But no encouragement?'

Well I do recall when I was still at school—I suppose I must have been fourteen or fifteen at the time—I wrote to him and said my aim was to become a great scientist, what should I do? And he wrote back and said, 'you should continue to work hard at school and pass your exams.'

'But why was your mother, who's not a scientist, so keen on science?'

Well, that's harder to say. I suppose she must have been deeply influenced by her father and her brother. She met my father through her brother—Haldane and my father, who was a lawyer, were school friends—so there had always been a jolly mixture. And it was certainly the way she was brought up, and the way she has been ever since.

'The three brothers all became biologists. Why did you choose biology, all three of you?'

Because that's what these previous generations in the family had done. People become biologists for positive and negative reasons. They become biologists because they're extremely interested and enthusiastic and exploratory, and that always takes one into biology because of its wonderful diversity of subject matter. People also go into it for negative reasons—basically at some point or another they've cracked up over mathematics. I certainly did. I could never have gone into physics because it was too deep for me.

'You don't think that it's something to do with the country house tradition? I'm struck how many grand, English families are connected with biology.'

I think you're absolutely right about that. A big country house has birds to pluck, fish to have their guts removed, cream to take off the milk: that's all biology. I did my share of that at home, and it certainly stimulated my interest in biology. But there was something else that seems more important to me in retrospect. Our house was filled with my parents' political and intellectual friends. I think the interest in biology is because of its breadth. Since childhood I've thought that biology had a great deal to do with the political positions which people take up regarding the past and future of mankind. The deep problems of population and growth, the ethical structure of human society, all that seems to me to be quite naturally a part of biology. I suppose it's also got something to do with physics, but physics is just that much remoter from human affairs. Biology is totally bound up with human affairs.

'I don't understand what you mean when you say that biology is intimately involved with politics.'

Well, you have to think back a little bit for that connection to be immediately apparaent. Nobody in the 1930s would have questioned that connection. The relationship between Fascist race theory and biology was transparent. The socialist countries rested their claim to excellence, in part, because they argued that socialism grows out of Darwinism.

'But are you saying that all these people went into biology because of their political commitment, rather than their interest in natural history?'

No, it was an argument from breadth I was using. Politics is part of that. The involvement that I mentioned, remember, was quite largely spurious in the form that I was talking about it, in the '30s. The socialist claims for a special relationship with science are spurious, and needless to say, the Fascist relationship with science was not only spurious, but utterly wicked. So all that's gone, and good riddance to bad rubbish, but the thinking through of that had its own interest at the time. Now there are other connections, but they're less immediate.

'But do you think biology does actually tell us anything about how we should live?'

Yes I do. I'd be hard put to say exactly what the message is, but I don't think science is anything very special there. I think it's to do with thinking clearly, testing ideas against reality, thinking quantitatively in a numerate fashion. I think if you add all that up together, oddly enough it does tell us a lot about how one ought to live.

'But biology itself?'

Well, I don't buy Lorenz lock, stock and barrel. I don't buy the notion that by studying aggression in sticklebacks you can learn all that there is to know about aggression in humans. But I do think that if you understand something about modern ethology, you will necessarily know more about human relationships. Perhaps not because of a logical connection, but because once you start thinking about conduct in a rational way, you don't stop at animals, you start thinking about your own species too.

'You don't think there's a danger, though, in extrapolating from the one to the other?'

Yes I do. I entirely understand that point. I have given lectures on altruism and insects, and the way their chromosomes are arranged and all that, and I think it's misleading to extrapolate directly. On the other hand, I think it's very important that one should learn to think in a straightforward way about altruism. Altruism is just as important as the bank rate, or how to make a Ford car. A lot of people are quite comfortable thinking through how to make a motor car, or why the bank rate goes up and down, but they're not at all comfortable about thinking through altruism, and I think that's the message of biology, that one should do both.

'It's somewhat curious that none of your sisters became scientists. That's true isn't it?'

Yes it is. Now why is that? Is it, do you think, that daughters see through their mothers, in a way that sons don't?

'Very possibly.'

I think that may be part of it.

'Because you had a very powerful mother. You still have a very powerful mother.'

Yes, yes. I think that sons ask less questions about their mothers than daughters do.

'Was it a competitive family?'

I wonder. Not terribly I think, no. It wasn't competitive in the sense that the family elders were in the habit of reprimanding the youngers for slackness, or wrong thinking, or anything like that. Not as far as I can remember. It was never as competitive as, let's say, any ordinary Japanese family is competitive at that age.

'There wasn't competition amongst the siblings then?'

No, not a lot. I think they tried to help one another. I think brothers often tend to help one another as, indeed, do sisters. I suppose there may have been a certain amount of boys ganging up against girls, and vice versa.

'Why did you choose immunology, having chosen biology?'

I just drifted into it. As an undergraduate student I fell under the spell of the great Peter Medawar. At the time I didn't realize immunology was what I was doing. Under his influence, I got deeply interested in the rejection of transplants. Indeed, at that time, in the late 1940s, it was actually unknown whether that was a bit of immunology or not. And indeed, immunology was a term which was hardly used. Immunology was something to do with antibodies. Nowadays the immune system represents a splendid body of information about a whole set of organs and tissues in the body, and is entirely comparable with, say, the nervous system. But in those days, immunology was, if anything, something to do with bacteriology, and didn't have anything to do with much of the rest of science. I was part of the generation who had the great privilege of bringing immunology in from the cold in that way, making it part of ordinary biology.

'What was the spell that Medawar cast on you?'

Well, you have to imagine a rather ordinary English lad—me—confronted with a brilliant, exotic scientist who seemed to be about 6′6″ tall, of great size and stature, of great charm, immensely communicative, an extraordinarily vivid personality.

'Do you think he exerted almost an undue influence because of this charisma that he had?'

Probably. Although I think it's difficult to stop 18-year-olds from developing, what should one call it, hero figures, crushes on teachers who teach them well. I think that's in their nature.

'Did you have a crush on Medawar?'

Well, I suppose so, in the sense that I became totally devoted, and even though it's now forty years later or so, I still dream about him from time to time.

'Did you copy his style of science?'

I copied everything about him. When I arrived in Oxford I wrote my letter 'e's in the way that everybody in English schools is taught to write 'e's, you know, with a sort of loop which comes up and round. I noticed he wrote his 'e's like the Greeks do, as 'ε', so I started doing that, and it still confuses bank managers, because I didn't manage to get all my 'e's exactly the same as his 'e's, so my writing's a bit confused still.

'Is your style of science similar to that of your brothers, because both of them are professional biologists?'

Fairly close. I suppose all biologists have a lot in common, and I think we're all representative, as you are yourself, of a generation which antedates the larger groups of scientists which have tended to grow since—'big' science, the 'team', all of that. We were brought up to be, and have remained, 'small' men in that sense.

'In a way, science is quite conservative, or it's perceived as being conservative. You come from quite a radical background; once again, your mother, and certainly your uncle, J. B. S. Haldane, was radical. How have you, as it were, adapted to this?'

That, again, is an interesting way of looking at things. There is a sense in which science *is* conservative, and science these days, especially in the context of universities, I think, genuinely is conservative. I think when the 1960s shook the humanities, the sciences drew back their skirts, waited for it all to blow over, and felt that it was no threat to them—that they could take a sort of comfortably patronizing line. Like a lot of other scientists, I didn't feel altogether comfortable about that at the time. I wish science was more deeply involved in the surge of human affairs, but it isn't. It stands apart from it. But remember, there's also a sense in which science is fairly radical, because there's no part of the scientific structure which stands absolutely secure. The whole point of the relationship between scientists who are trying to understand how nature works, and nature itself, is that nature can always surprise them. Science is an articulated structure of hypotheses, as Peter Medawar said, and the better you are as a scientist, the more you'll manage to shake that structure.

'But the formal structure of British science is pretty conservative, have you found?'

Ah, you don't mean the logical structure, you mean the social structure, with professors and heads of institutes and all that, the hierarchical nature of science. Well, yes and no. I think where it matters, at the chalk face of your teaching, or at the laboratory bench, or when confronted with a group of animals, or a group of plants, that hierarchy means absolutely nothing. The whole point of science is that the most junior participant is entitled to discover secrets of nature, and I think that's why it's all much more acceptable than are the hierarchies which are equally to be found in the legal profession, or the army, or whatever, and there's no way out.

'But I think this attitude must be peculiar to the way *you* run science, because many departments are very hierarchical.'

Well, do you think that any department which genuinely practises scientific research is hierarchical in that sense?

'Interesting question. I don't know.'

I think there are lots of organizations which purport to practise science, and if they turn out to be hierarchical, then you can be pretty sure that they're not doing anything worthwhile anyway.

'But how do you run your department then?'

I don't know that I do much running. There isn't much called for. Actually, I suppose that's exaggerating a bit. There are some very important things which people who are older and more senior in the hierarchy do. One is to keep making judgements about younger people. They have to tell department X, where some young colleague has asked for a job, whether the young colleague in question is capable or not, and that may be the most important thing they do. In principle you can learn that by reading the young colleague's papers. In practice, it is important to get opinions from older people, too. And I think there's another important thing which older people can do, and that is to help, aid and comfort the young, who are afflicted with all the problems of life just because they are so young.

'Now there's a slight contradiction. Not in what you've said, but in what I know about you. On the one hand, I know you to be very kind. On the other hand, I know you to be intellectually extremely rigorous and ruthless, perhaps almost to the point of cruelty, or am I being unfair?'

No I think that's true and a terrific compliment. I think that's disconcerting for very young people. The younger you are the more disconcerting it is to find that you're made welcome in some place—and I certainly do try and make people welcome—but presently to find that you're ignored because you're quite clearly regarded as too stupid to be worth listening to. I'm afraid that there are people who get into that position, and they would be far better to move out, and they do move out. Science is very selective, isn't it, all the way through. It's not true that if people get trained in the right way, then they will do all right. I think there's an element of training, but a great deal of the educational process is selection, isn't it. People are given the opportunity to train themselves, and if they succeed, then they can go on to the next stage, and if they don't, even if nobody tells them in so many words, they'd do better to get out of it.

'What do you regard as your skills as a scientist?'

They're partly to do with persistence. On these sorts of scales from one to ten which one might mark oneself on, I rate persistence high. It's also very important to know when to stop doing something, and I suppose I score rather lower on that. Probably I go on doing things perhaps over-persistently. I suppose being alert to current developments is very important, and I think I'm quite good at that. I've been fortunate, I think, in that people, actually quite early in their careers, can get into a sort of position in science where they go round to lots of meetings and they talk to people, and they start forming a wide circle of scientific friends, and benefiting from that enormously.

'So by having this wide set of contacts you've been able to quickly introduce new discoveries. Is that what you mean by that?'

Yes, yes. Not just new discoveries. I think I spend, everybody spends, but I'm very conscious of spending, a lot of my time trying to imitate or learn from other people, from the activities of one's peers.

'What about your experimental skills?'

Of course as time passes one becomes more skilful in some things than most other people, and it's also true that as one gets older, skills which were ubiquitous among the people who taught me, are now becoming incredibly rare. I can pick up mice with one hand, and keep a syringe in the other and inject them, so that I can inject 500 mice in an hour. Because most people do things in test tubes now, they can't do that.

'What do you feel about working with mice?'

I'm well aware that in a way my sensibilities are dulled. That is, I will do things to animals which the more squeamish would not do. That is not the same thing as saying that I'd be cruel to them, because I'm convinced that most animals object to being handled incompetently, or slowly or being fiddled with. I think they object much less to being picked up firmly by the scruff of their necks. I don't really think they mind that very much.

'Do you actually care about the mice?'

Of course I care about them, yes. Do I try to minimize cruelty to them? Yes, of course I do.

'Do you actually like doing the experiments with these skills?'

Oh yes, oh yes. I mean no competent surgeon can not enjoy surgery. Anybody who can handle animals properly, enjoys doing so. Any farmer does. The pleasure in handling a mouse properly is no different from the pleasure taken in properly administering a pill to a cow, as any farmer does.

'What then would you say was your real pleasure in doing research, or doing science?'

The conduct of an experiment from beginning to end, an experiment which starts with, again I quote Peter Medawar, 'the act of creation'. It's absolutely true. Nobody who has ever thought of a good experiment will ever mistake what thinking of a good experiment is. Not all experiments that you think of are, of course, good experiments, but thinking of a good experiment is just wonderful, eureka! It's fantastic. Then there's a lot of hard work; then data which either support, or disprove, the hypothesis which came in the act of creation, come forth; and if they support the hypothesis, that's absolutely fantastic. And, of course, it's very nice at the end to be able to tell one's friends about it. One looks forward to doing that.

'Do you care about the social value of what you discover?'

A bit. If one does, as I do, or have done, experiments which bear on organ transplantation, or which bear on leprosy, that carries with it a certain satisfaction. I must say that I'm also very well aware that one of the many traps which scientists fall into is doing science which is useful without being good science, because I think that's a non-category. I think you can deceive yourself that you're doing useful science without it being good or beautiful science, but that just doesn't exist. It's got to be good and beautiful science. If it's immediately useful as well, that's great. I think very little science which is true and beautiful ends up, in the long run, not also being useful.

'So you don't think that there's some gap, an unfortunate gap, between the practice of science and the real world outside it?'

Well I suppose, again, that as one becomes older and wiser, that gap is more evident. Let's take as an example a leprosy vaccine. You know that I am trying to participate in the development of a leprosy vaccine. I think when one starts in science one can see no gap between immunizing a mouse successfully against some bit out of a leprosy bacillus, and the development of leprosy vaccine. I can now see an enormous gap. I can see a gap which is to do with money and all the problems of implementing things in developing countries, I can see all the other kinds of inputs which are needed for that kind of medical advance.

'When you talk about a beautiful idea, what are you talking about?'

Well, you know I'm tempted to remind you about Bridges' definition of good poetry. It's what makes the bristles on your chin stand up when you shave in the morning. I suppose a beautiful idea in science is something which is simple and comprehensive and which makes a prediction which is really unexpected.

'When one has this experience of beauty, is it weekly, monthly? I'm trying to get a sense of frequency.'

I think one's doing extremely well if one has a good idea once every, let's say, six months, twice a year.

'That's doing very well I would say. I think you're doing very well. And the imaginative process that gives you these beautiful ideas?'

Ah well, there you're asking a very deep question about the human spirit which I don't think anybody can possibly answer. It's certainly not good for sleeping at night. Good ideas are very closely connected with what keeps one awake. The bath is a bit of an illusion. I've never had a good idea in the bath. Have you ever had a good idea in the bath?

'Excellent, yes.'

You have had a good idea in the bath?

'Yes.'

How lovely, [laughter] oh I do envy you that. No I've never had a good idea in the bath. Perhaps I haven't taken enough baths, or stayed in for long enough.

'Where do you get your good ideas then?'

Well I think in retrospect one always knows where good ideas come from. They come from logical relationships which one hadn't perceived before. Once they're there, they're crashingly obvious.

'But do you set out consciously to think of an idea? In other words, your persistence that you spoke about, do you work at a problem in order to get a solution, or do you sit around waiting for inspiration?'

I think one knows some ways of not having good ideas. You know a long rest, a nice summer holiday, that's not a great way of having good ideas. Good ideas come from exercising the mind.

'In fact you list in *Who's Who*, no recreations.'

Yes. Well I suppose it's not true that I don't enjoy other things. But I must say that there are certain ways in which I'm not in sympathy with the whole point of this programme, and that is that I think that the cult of the personality is profoundly unsympathetic. I think it's unsympathetic to science and I just don't like it myself either. Perhaps I've had too much of it. All the things which we were talking about earlier. Perhaps I've been too much overawed in the past. I think the whole notion that great men should be treated like pop stars, and one should know all about them, is absolute baloney. I think that what's interesting about scientists is the science that they do, and not the recreations that they have.

'You don't think it's important that the public should perceive scientists as actually having recreations? Because there is a tremendous, I think, public misconception about the very nature of scientific work and what scientists are like?'

But do you think that the general public is so silly as to think that scientists don't do other things. It's just that if a scientist is a gardener, or a bee keeper or whatever it is, he's not going to be a better gardener, or a better bee keeper than somebody who is really good at gardening or bee keeping. I suppose once in a generation there's an exceptionally good scientist who is also an exceptionally good bee keeper, but by and large if I wanted to hear about bee keeping I'd go and talk to people who were really good at bee keeping, and if one of them happened to be a scientist it would amaze me.

'What about the next generation? You can't avoid the fact that you come from a scientific family. At least one of your sons, Tim, is an excellent scientist. Does that matter to you, handing on to the next generation the same skills, as it were, that you inherited?'

I'm as pleased as punch to have a child, at least one—I've hopes of another as well—who will be a good scientist. I'm not worried about the future of science in general. Thatcherite Government has a lot to answer for. It's engaged in shrinking the State, and poor old Science is getting a bit of stick in the course of doing that. But in the long run I think that science and the progress of science totally dwarfs the activities of one particular lot of ministers. Who cares what they think? In the eyes of history they're going to be but a tiny footnote, whereas science is part of the great wave of the future.

Index